ROAD VEHICLE AERODYNAMIC DESIGN

ROAD VEHICLE AERODYNAMIC DESIGN

AN INTRODUCTION

R. H. BARNARD PhD, CEng, FRAeS

Principal Lecturer in Mechanical, Aerospace and Automotive
Engineering, University of Hertfordshire

LONGMAN

Addison Wesley Longman Limited
Edinburgh Gate, Harlow
Essex CM20 2JE, England
and Associated Companies throughout the world

First published 1996

British Library Cataloguing in Publication Data
A catalogue entry for this title is available from the British Library

ISBN 0-582-24522-2

Set by 32 in Ehrhardt 10/12 and Futura
Printed in Great Britain by Henry Ling Ltd., at the Dorset Press, Dorchester, Dorset.

CONTENTS

ABBREVIATIONS

bhp	brake horse power
CFD	computational fluid dynamics
coe	cab over engine
dB	decibel
EPA	Environmental Protection Agency (USA)
ESDU	Engineering Science Data Unit (UK)
km/h	kilometres per hour
MIRA	Motor Industry Research Association (UK)
mph	miles per hour
NACA	National Advisory Committee on Aeronautics (USA)
NMI	National Maritime Institute (UK)
NRC	National Research Council (Canada)
psi	pounds per square inch
RAeS	The Royal Aeronautical Society (UK)
rms	the square root of the mean of the squares
SAE	Society of Automotive Engineers (USA)
SPL	sound pressure level

NOMENCLATURE

A	area (usually projected frontal area)
A_c	frontal area of radiator core
A_∞	area of stream tube at upstream infinity
b	breadth
C_D	drag coefficient
C_L	lift coefficient
C_M	pitching moment coefficient
C_R	rolling moment coefficient
C_S	side force coefficient
C_Y	yawing moment coefficient
d	diameter
D	drag
F_r	rolling resistance
F_t	tractive force
g	gravity constant
I_u	turbulence intensity
k	a constant
$k_c(\max)$	maximum lateral (tyre) adhesion or cornering coefficient
k_r	rolling resistance coefficient
k_p	pressure loss coefficient in radiator core
$k_s(\max)$	maximum value of longitudinal (tyre) slip coefficient
l	vehicle length
L	lift
L_d	aerodynamic down force (negative lift)
P_a	power available
r	radius of curve
Re	Reynolds number
S	wind-tunnel working section area
S_c	centripetal force
t	time (also track width)
u	instantaneous value of longitudinal velocity above mean value
U	horizontal component of velocity

V	speed
W	weight
x	longitudinal distance from origin
y	lateral (sideways) distance from origin
z	vertical distance from origin
α	index in the velocity profile power law
β	underbody diffuser angle
Δ	denotes a change
θ	backlight (rear screen slope angle) to horizontal (also angle of inclination of road)
θ_{eff}	slope angle between rear roofline and back of boot (trunk) top
μ	air dynamic viscosity coefficient and sliding friction coefficient
ρ	air density
ψ	yaw angle: angle between car axis and relative wind direction
$\bar{}$	(overbar) denotes a time mean value

PREFACE

The study of road vehicle aerodynamics is quite different from that of aircraft aerodynamics. The subject does not lend itself to traditional methods of mathematical analysis, and there are no equivalents of the simple formulae used in earlier days for determining the lift and drag of an aircraft from its geometry. The road vehicle aerodynamic designer has very few mathematical tools. A few simple equations have been included here, particularly in the chapters on basic aerodynamics and performance, but the book concentrates mainly on explanations of the physical principles and on the methods of experimental measurement that can be used. A description is also given of some of the many experimental studies that have been conducted, mostly in the last two decades. Armed with this information, the reader should be able to make a reasonable attempt at a low-drag conceptual design, and should also be able to assess the reasons for aerodynamic defects in an existing shape.

Most major motor manufacturers now possess a large wind-tunnel, and employ one or more aerodynamic specialists. There is currently no common view amongst companies as to where these specialists belong; they may be part of the styling department or the engineering department, or even in a separate unit. Vehicle aerodynamics is, however, not just the preserve of specialists; anyone concerned with vehicle styling or bodywork design should have a basic understanding of aerodynamic principles, so that their application becomes an inherent part of the design process, and is not just an added afterthought.

This is not intended as a history of the subject, and historical material is mainly restricted to cases where the development helps to explain the principles involved, racing car aerodynamics being a particular example. The book is intended only as an introduction, but numerous references are given that will take the reader deeper into specialist topics.

R. H. Barnard

ACKNOWLEDGEMENTS

The author wishes to thank the following for their help in providing diagrams, photographs and other material during the preparation of this book:

Ms K. Ager, LAT Photographic; Paul Blackmore, Building Research Establishment; Dr Brennenstuhl, Volkswagen; Mr H. Calton, Aston Martin Lagonda Ltd; Mr G. Carr, Dr J. Graysmith and Mr K. Read, MIRA; Dr K. Garry, Cranfield University; Dr J. Howell, Rover Group Ltd; Prof J. Katz, San Diego State University; Ms F. Loader, Porsche Cars Great Britain Ltd; Mr Nayler, the Royal Aeronautical Society; Terry Newman, Dr K. Mallone and Dr S. Wakes, University of Hertfordshire; Dr M. Ramnefors, Volvo Data AB; Prof. Rolsenblat, Fluid Dynamics International, Inc.; and Prof. Dr.-Ing. H. Zingel, Fachhochschule Hamburg.

A number of students of the University of Hertfordshire have provided useful data, including: R. Blaylock, J. Comer, I. Evans, R. Fisher, B. Gilmour, P. Monaghan, M. Naji, M. Savill, and H. Vaughan.

Particular thanks are due to Dr C. Baker, Nottingham University, for his helpful comments and suggestions on the topic of cross-wind effects, and to Sue Attwood who turned my untidy drawings into works of art.

The publishers are grateful to the following for permission to reproduce copyright material:

Rover Group Limited for Figs 1.14, 6.33 & 11.14; R. Davies for Fig. 2.2; Volkswagen AG and Deutsche Forchungsanstalt für Luft- und Raumfahrt for Figs 4.2, 4.5 & 4.16; Jeff Harkham for Fig. 4.12; Terry Newman for Fig. 4.29; Dr K. Garry for Figs 4.37–40; R. G. Dowson for Fig. 5.2; the Motor Industry Research Association for Figs 5.6, 8.4 & 12.5; Prof. J. Katz for Figs 6.23 & 6.24; LAT Photographic for Fig. 6.25; Aston Martin Lagonda Ltd for Fig. 9.6; Porsche AG for Fig. 9.7; Dr C. Baker for Fig. 10.11; the Building Research Establishment for Fig. 11.10 (© Crown copyright); Volvo Data AB for Figs 12.6 & 12.9; Dr S. Wakes for Fig. 12.8; Dr K. Mallone for Fig. 12.10.

CHAPTER 1

UNDERSTANDING AIR FLOWS

AERODYNAMIC ANALYSIS AND AERODYNAMIC DESIGN

As mentioned in the preface, an important feature of the subject of road vehicle aerodynamics is that it does not lend itself readily to mathematical analysis; there are no straightforward methods for predicting how air will flow around a given vehicle shape. This is surprising in view of the sophistication and variety of methods of analysis available for aircraft aerodynamics. The analytical difficulties arise from the fact that the flow around a road vehicle is highly three-dimensional, the air does not follow the contours of the body everywhere, and there is nearly always an unsteady wake. The problems of analysis are described in more detail in the last chapter, where it is also explained why vehicle aerodynamics presents a severe challenge even to the powerful computational fluid mechanics methods now available.

This book is not so much concerned with the analysis of flows, but rather with the problem of how to design vehicle shapes that produce desirable flow characteristics. At present, nearly all aerodynamic design for road vehicles relies on a combination of experimental results, experience, and a physical understanding of the way that air flows behave. Much aerodynamic development involves trial and error experiments using wind-tunnel models (Fig. 1.1). This does not mean, however, that the subject is totally free of mathematics. It is still necessary to have an analytical basis for the methods used to treat the experimental data, to predict performance, and to relate wind-tunnel results to full-scale behaviour.

It is assumed that most readers will already have at least some very basic knowledge of fluid mechanics, but because of the importance of gaining a proper understanding of the physics of air flows, some of the basic principles are restated in this chapter, and their relevance to vehicle aerodynamics is shown. For completeness, a few essential definitions are also included. This may seem a rather dull way to start on an inherently interesting subject, but the chapter will at least provide a point of reference if unfamiliar terms and concepts are encountered elsewhere in the book.

Fig. 1.1 Clay wind-tunnel models are often used for the early stages of aerodynamic developments. As the design progresses, more accurate and detailed models are used. In this example, wool tufts have been taped to the surface to provide a simple means of flow visualization.

AIR DENSITY AND VISCOSITY VALUES FOR ROAD VEHICLE AERODYNAMICS

The air density (mass/unit volume) is conventionally denoted by the symbol ρ, and the standard atmospheric sea-level value is 1.226 kg/m^3. There is an unfortunate tendency to think that the atmospheric density does not vary very much on the Earth's surface, but that is quite wrong; for a given atmospheric pressure, the density is inversely proportional to the (absolute) temperature, and that can vary from as low as 213 Kelvin (−60 °C) in the Arctic to some 323 Kelvin (50 °C) at the equator. The density also varies with height; on a 3000 m (10 000 ft) high mountain pass, the density will be only 75 per cent of the sea-level value. Aerodynamic forces such as lift and drag are directly proportional to the air density, so variations in atmospheric air density can have a large effect on vehicle performance; calculations are however almost invariably based on the standard sea-level value. At speeds of less than 400 km/h (250 mph) the air density does not vary locally around the vehicle body to any significant extent, and in any analysis, the air is treated as being incompressible.

Viscosity is, in simple terms, the stickiness of a fluid (liquid or gas). The viscosity coefficient μ gives a measure of the force required to shear a layer of fluid at a given rate. (The mathematical definition will be found in any elementary textbook on fluid mechanics). Air is around 64 times less viscous than water, and in the early days of the study of the mechanics of fluids, it was therefore thought that the viscosity of the air would have a negligible effect on the way that it flows around an object. This was soon found to be completely untrue; viscosity has a strong influence on air flow behaviour, and without viscosity there would be neither lift nor drag forces. The standard sea-level value for air (dynamic) viscosity coefficient is 1.78×10^{-5} kg/m s.

MOVING VEHICLES AND MOVING AIR

Aerodynamicists nearly always prefer to think in terms of what happens when air flows past a stationary vehicle, rather than the real situation where the vehicle moves through the air. This is because it is generally much easier to understand and describe what happens in the former case. Fortunately, in terms of the forces and flow features, it makes no difference whether air is blown past a stationary object in a wind-tunnel or the object is moved at the same speed through stationary air; it is the *relative* speed that matters. A car travelling through still air at 20 m/s will experience exactly the same aerodynamic forces as a stationary car subjected to a head-on 20 m/s wind. For road vehicles there is one slight complication in that there is also relative motion between the road and the vehicle, so in the stationary-vehicle case, it is necessary to imagine that both the air and the road are moving past the vehicle. In the case of a vehicle travelling along a road in windy conditions, the relative air velocity and the relative road velocity are not the same, and this results in some experimental and analytical problems, as will be described later.

FREE-STREAM SPEED

When a vehicle is placed in a wind-tunnel, the relative speed of the air stream (away from local changes caused by the presence of the vehicle) is referred to as the **free-stream speed**. For the case of a vehicle driving along a road in windless conditions, the corresponding relative speed is simply the driving speed of the vehicle.

THE CONNECTION BETWEEN AIR SPEED AND PRESSURE - THE BERNOULLI RELATIONSHIP

When air flows from a region of high pressure to one at a lower pressure, the pressure difference provides a force in the direction of flow, and the air therefore accelerates. Conversely, flow from a low pressure to a higher one results in a

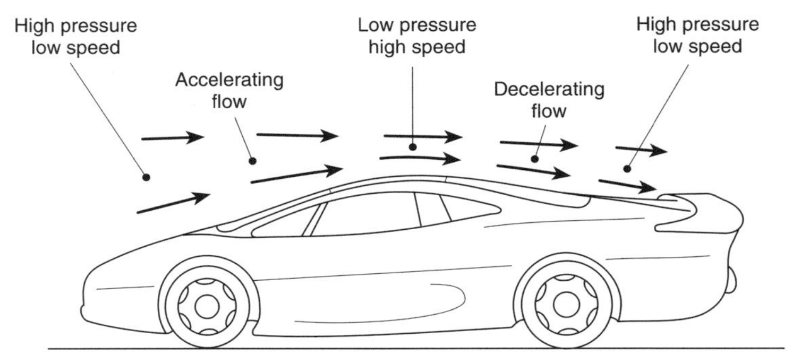

Fig. 1.2 Pressure and speed
The air accelerates when flowing from a high pressure to a low one, and slows down when flowing from a low pressure to a high one.

decrease in speed. As illustrated in Fig. 1.2, regions of high pressure are therefore associated with low speeds, and areas of low pressure are associated with high speeds (in the absence of friction effects).

Although road vehicle aerodynamics may not generally lend itself to simple mathematical analysis, there is one relationship that is absolutely fundamental to the study of air flows, and that is the Bernoulli equation. For low-speed flows, this equation gives the relationship between air speed and pressure. It can be written in several different ways, but aerodynamicists normally prefer it in the form given below:

$$\text{(pressure)} + \tfrac{1}{2}\text{(density)} \times \text{(speed)}^2 \text{ is constant}$$

or in mathematical symbols

$$p + \tfrac{1}{2}\rho V^2 = \text{constant}$$

where p is the pressure, ρ is the density, and V is the speed.

It may be seen that this relationship fits the behaviour of the air as described above, in that an *increase* in pressure must be accompanied by a *decrease* in speed, and vice versa.

Readers who have already encountered Bernoulli's equation in fluid mechanics may recognize that the above expression is just a version in which the height term has been ignored; changes in this term are usually insignificant in automotive air flows.

Bernoulli's equation is derived using the assumption that the air density does not change with pressure. This is not really valid, but at the speeds at which even racing cars travel, the equation will give very accurate results. Only absolute record-breaking vehicles travel at speeds which are high enough to make the equation invalid due to compressibility effects.

STATIC AND DYNAMIC PRESSURE

The first term in the above equation is just the air pressure at any point, and is also known as the **static pressure**. The second term $\frac{1}{2}\rho V^2$ is called the **dynamic pressure**. Although it has the same units as pressure, dynamic pressure actually represents the kinetic energy of a unit volume (m^3 or ft^3) of air. Aerodynamic forces such as lift and drag are directly dependent on the dynamic pressure. It is therefore a term that crops up frequently:

$$\text{dynamic pressure} = \tfrac{1}{2}\rho V^2$$

Since the *local* flow speed varies around the surface of a vehicle, the associated local static and dynamic pressures will also vary with position.

It is often convenient to refer to the dynamic and static pressures in the free stream: that is, away from the influence of the vehicle, and the terms **free-stream dynamic** and **free-stream static** pressure are used frequently. For a vehicle travelling along a road, the free-stream static pressure is simply the atmospheric pressure, and the dynamic pressure is

$$\tfrac{1}{2}\rho \times (\text{driving speed})^2$$

STAGNATION PRESSURE

From Bernoulli's equation it can be seen that at any point in the flow where the speed is zero (the lowest possible value), the pressure must reach its maximum value. Because this maximum pressure is associated with stagnant conditions (zero flow speed), it is known as the **stagnation pressure**. In the context of cooling air flows, and engine intakes, the stagnation pressure is commonly referred to as the **ram air pressure**.

The similarity of the words static and stagnation leads to some confusion. Static pressure, as defined previously, is simply the pressure p in the flow. The pressure only reaches the stagnation value at a point where the flow speed is zero. Energy losses due to friction produce a loss in stagnation pressure.

PRESSURE COEFFICIENT

The difference between the local static pressure at any point in a flow and the static pressure in the free stream depends directly on the dynamic pressure of the free stream. Therefore, the ratio

$$\frac{\text{local pressure} - \text{free-stream static pressure}}{\text{free-stream dynamic pressure}}$$

remains constant at all speeds. This ratio is known as the pressure coefficient and is denoted by C_p.

$$C_p = \frac{p - p_\infty}{\frac{1}{2}\rho V_\infty{}^2}$$

where p is the local static pressure

p_∞ is the free-stream static pressure

V_∞ is the free-stream speed (the driving speed in still air)

Note that the subscript ∞ is frequently used to denote free-stream conditions.

When describing how the pressure varies around a vehicle, it is much more convenient to use the pressure coefficient rather than the actual pressure, because the coefficient will not alter with the vehicle speed. Knowing the value of the pressure coefficient at a point, it is a simple matter to calculate the pressure there at any free-steam or driving speed, using the above relationship.

THE DIFFERENCE BETWEEN SOLID PARTICLE MOVEMENT AND FLUID FLOW

An important step in understanding air flows is to appreciate that air and other fluids do not behave in the same way as solid objects or even streams of solid particles; air molecules do not impact on the front of a vehicle, they flow over it. Although air is composed of individual molecules, they do not behave independently; a change in the pressure at any point to some extent affects the pressure and hence speed everywhere else. The influence of any change propagates at the speed of sound, which is around 340 m/s at sea level. Since all but record-breaking cars travel much slower than the speed of sound, this high speed of propagation means that pressure disturbances can be be transmitted upstream into the oncoming flow. The forward propagation explains why modifying the rear-end shape of a vehicle not only changes the local flow speed and pressure, but alters the whole flow field around the vehicle, even at the very front.

STREAMLINES

Figure 1.3 shows the streamlines of the flow around the centre-line of a vehicle. Streamlines are defined as imaginary lines across which there is no flow. If the flow is steady they also indicate the instantaneous direction of the flow and the path that an air particle would follow. For most types of road vehicle there are usually regions of unsteady flow. The instantaneous flow direction can become very erratic in these areas, and the shape of the streamlines changes rapidly with time, but it is still possible to indicate the *average* flow direction by a form of streamline which denotes a line across which the average flow is zero.

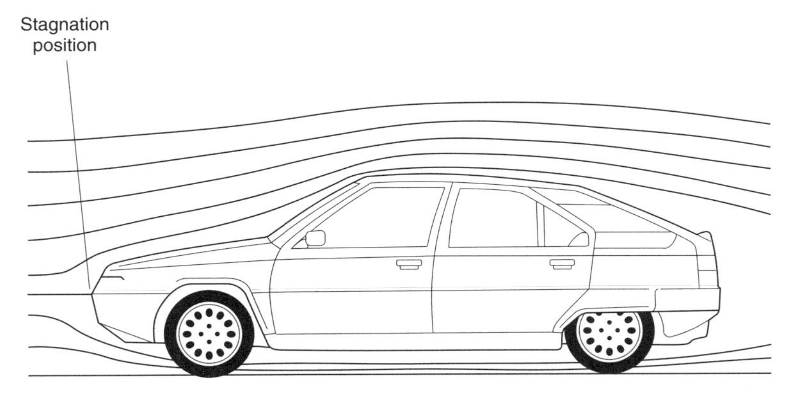

Fig. 1.3 Streamline patterns around the centre-section of a vehicle.

STREAMLINED SHAPES AND ATTACHED FLOW

The term streamlined is colloquially applied rather loosely to any vehicle with flowing curved lines. In more scientific terms a streamlined shape can be conveniently defined as one where the air flow follows the contours. In this case, the flow is said to be **attached**. Note that this definition of a streamlined shape is not a universally accepted one, but merely a convenience for present purposes. In Fig. 1.4 two shapes are shown, a 'streamlined' one where the flow stays attached, and a 'bluff' one where the flow separates from the surface near the front.

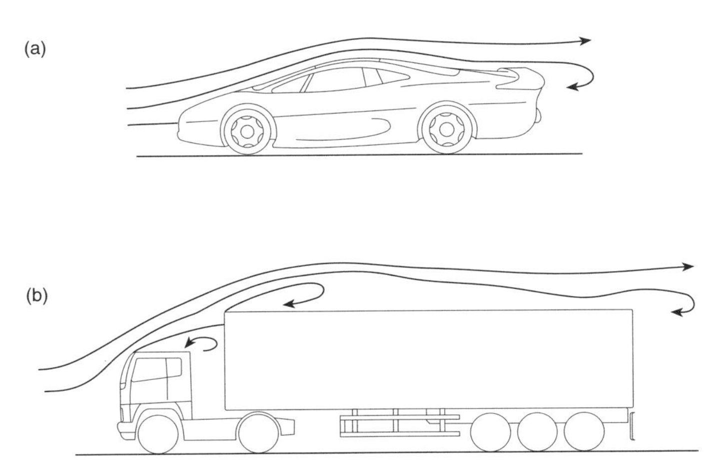

Fig. 1.4 (a) A streamlined vehicle shape – the flow follows the contours of the body up to the rear end; (b) a bluff vehicle shape – the flow separates from the top of the cab and from the front edge of the trailer.

The closeness of streamlines gives an indication of the local speed; where streamlines converge, the air is funnelled through at increased speed. It is therefore possible to determine the pressure and speed around a vehicle by photographing streamline patterns, and measuring the distance between the lines.

STAGNATION REGIONS

An interesting aspect of the flow shown in Fig. 1.3 is the behaviour of the air on the dividing streamline – the streamline that divides the flow that goes over the top of the vehicle from that which passes under it. At the point where the air strikes the body, it momentarily comes to rest, and the position where it occurs is called a stagnation position. At this point, the air pressure will have its stagnation value: the maximum that it can attain anywhere in the flow.

It is always rather bad practice to consider two-dimensional views of air flows around vehicles, because the flow is characteristically highly three-dimensional. Figure 1.5 shows a three-dimensional view. The streamlines have to be replaced by stream tubes. These are difficult to represent on a flat drawing, and part of the tubes has been cut away for clarity. At the rear of the vehicle, the flow pattern becomes too complex und unsteady to be represented by continuous stream tubes.

THE BOUNDARY LAYER

An important feature of the flow past a vehicle is that the air *appears* to stick to the surface. Right next to the surface there is no measurable relative motion. You may have noticed that loose dust particles are not blown off a car's surface even at high

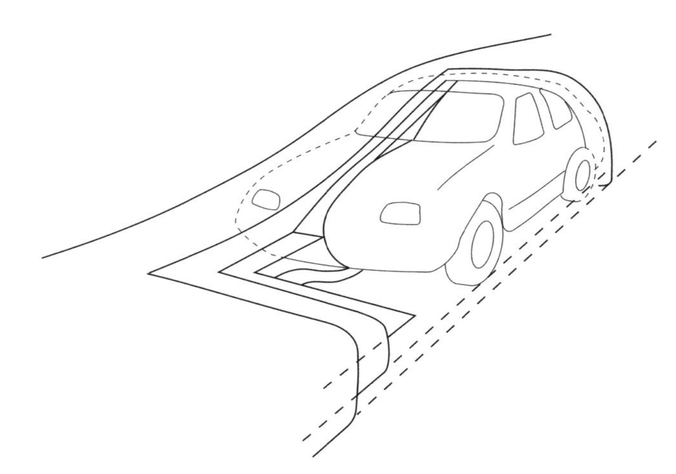

Fig. 1.5 Three-dimensional stream tubes
The tubes have been cut away to reveal the flow patterns.

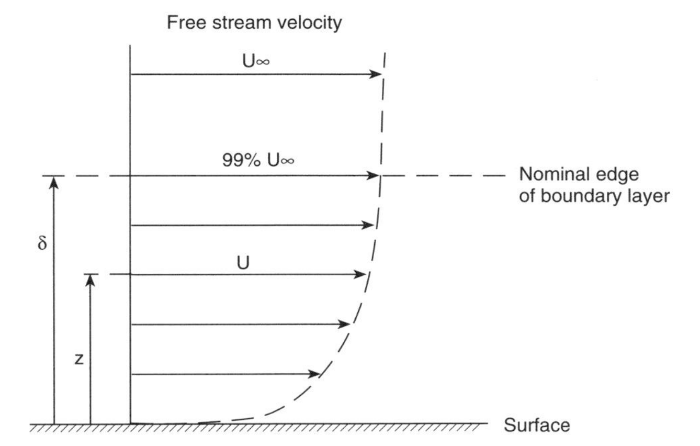

Fig. 1.6 The variation of velocity in a boundary layer
The profile shown is that for a turbulent boundary layer above a smooth flat plate following the law $U/U_{\infty} = (z/\delta)^{1/7}$.

speeds. Individual molecules do not actually physically stick, they move around randomly, but their average velocity component parallel to the surface is zero. The relative velocity of the air flow increases rapidly with distance away from the surface, as illustrated in Fig. 1.6, and only a thin layer is slowed down by the presence of the surface. This layer is known as the boundary layer.

The thickness of the boundary layer grows with distance from the front of the vehicle, but does not exceed more than a few centimetres on a car travelling at normal open-road speeds. Despite the thinness of this layer, it holds the key to understanding how air flows around a vehicle, and how the lift and drag forces are generated.

LAMINAR AND TURBULENT BOUNDARY LAYERS

Figure 1.7 shows how the boundary layer grows on the surface of a vehicle. There are two distinctive types of boundary layer flow. Near the front edge, the air flows smoothly with no turbulent perturbations, and appears to behave rather like a stack

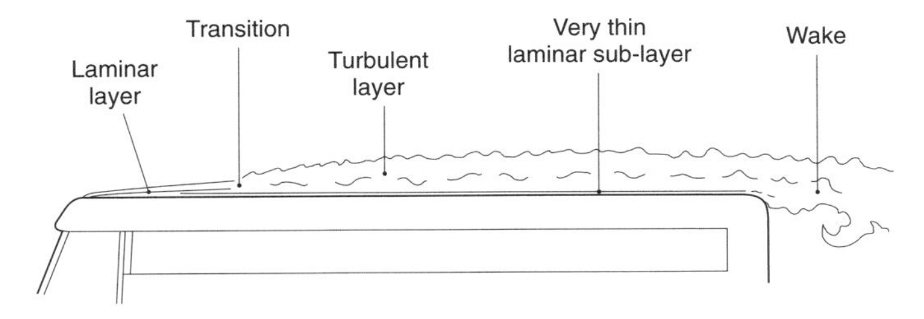

Fig. 1.7 Boundary layer growth on the roof of a bus
Note that the thickness of the layer has been greatly exaggerated.

of flat sheets or laminae sliding over each other with friction, the outer ones moving faster than the inner ones. This type of flow is therefore called **laminar flow**. Further along, as indicated in Fig. 1.7, there is a sudden change or **transition** to a turbulent type in which random motion is superimposed on the average flow.

It is important to distinguish between turbulent boundary layer flow and separated flow. In a turbulent boundary layer, the flow is still 'streamlined' in the sense that it follows the contours of the body. The turbulent motions are of very small scale. A separated flow does not follow the contours of the body.

The two types of boundary layer flow have important differences in their properties, and these can be exploited in aerodynamic design. In simple terms, the main practical effects are that the laminar layer produces less drag due to friction with the surface, but in a turbulent boundary layer, the flow is more likely to follow the contours of the body, which usually results in lower 'pressure drag' as described in Chapter 2.

To understand why these two types of boundary layer behave differently, it is necessary to look at their structure in a little more detail.

HOW THE BOUNDARY LAYERS FORM

In a laminar layer, molecules from the slow-moving air near the surface mix and collide with those further out, tending to slow down more of the flow. The slowing effect produced by the surface thus spreads outwards, and the region affected, the boundary layer, becomes progressively thicker along the direction of flow. At some distance along the boundary, an instability develops, and there is a **transition** to a turbulent motion, as described previously.

In the turbulent boundary layer, eddies form that are relatively large compared with molecules, and the slowing-down process involves a rapid mixing of fast- and slow-moving masses of air. The turbulent eddies extend the influence further outwards from the surface, so that at transition, the boundary layer thickens. Very close to the surface in a turbulent boundary layer there is a thin sublayer of laminar flow.

The important difference between the flow mechanisms in the two layers is that in the laminar type, the influence of the surface is transmitted outward mainly by a process of molecular impacts, whereas in the turbulent type the influence is spread by turbulent mixing. The turbulent mixing is a much more rapid process.

SURFACE FRICTION DRAG

Just as the surface slows down the relative motion of the air, so will the air try to drag the surface along with the flow. The process appears similar to the friction between solid surfaces, and is known as viscous friction. It is the process by which **surface friction drag** is produced.

The surface friction drag force depends on the rate at which the layers of air right next to the surface are trying to slide relative to each other. In the case of the laminar layer, the relative speed decreases steadily through the layer following a parabolic law. In the turbulent layer, however, air from the outer edge is continually mixed in with the slower moving air, so that the average speed close to the surface rises rapidly with distance from the surface, thus producing a greater amount of drag for a given thickness of layer.

On most current road vehicles travelling at the maximum legal speed, the laminar boundary layer does not extend for much more than about 30 cm (1 ft) from the front. It would require a radical change of styling to produce significantly larger amounts of laminar boundary layer. As described in later chapters, however, small wind-tunnel models can have a much greater proportion of laminar boundary layer.

FLOW SEPARATION

Figure 1.8 shows the flow over the top of two vehicles. The flow speed will be high near the top of the front screen at A, and the pressure correspondingly low. The pressure then gradually rises again as the flow speed decreases. This means that the air has to travel from a low to a high pressure, which it can do by slowing down and losing some of its kinetic energy. The situation can be likened to that of a car coasting up a hill, which is possible as long as it is travelling fast enough at the bottom. Close to the surface, in the boundary layer, some of the available energy is dissipated in friction, and the air cannot return to its original *free-stream* starting conditions of speed and pressure, just as friction will prevent a

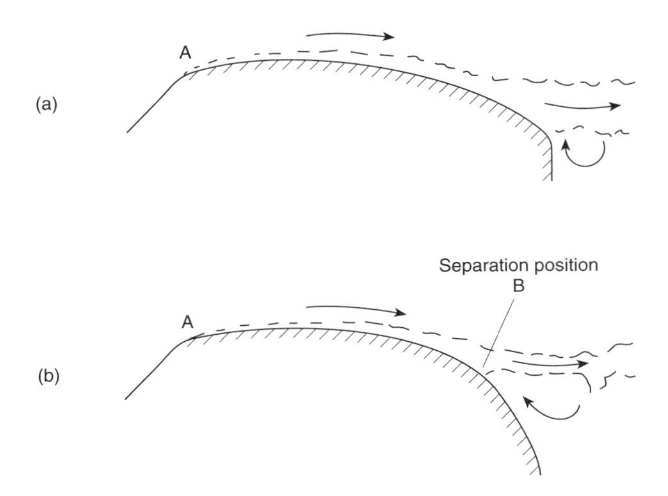

Fig. 1.8 Adverse pressure gradients and flow separation
The more strongly adverse pressure gradient in (b) causes the flow to separate at B before reaching the rear of the vehicle.

vehicle from coasting up a hill the same height as the one that it has just coasted down. If the increase in pressure is gradual as in Fig. 1.8(a), the process of turbulent mixing or molecular impacts allows the outer layers to effectively pull the inner ones along. Energy is fed in from the fast-moving air outside the layer. If the rate of increase of pressure is too great, however, as in Fig. 1.8(b), the mixing process will be too slow to keep the lower part of the layer moving. When this happens, the boundary layer flow stops following the contours of the surface and *separates* away, as shown at position B in Fig. 1.8(b). Air particles downstream of the separation position tend to move towards the lower pressure in the reverse direction to the main flow. There is still a layer of air in which there is a rapid shearing effect, but as it is no longer attached to the boundary, it is called a separated shear layer.

Unless the rear of the vehicle tapers down to a long tail, which is rarely the case, the flow will always separate at the rear as in Fig. 1.8(a), if not before, as in Fig. 1.8(b).

FAVOURABLE AND UNFAVOURABLE CONDITIONS

As described above, separation tends to occur when the air flows from a low pressure to a high one. This is therefore known as an **adverse pressure gradient**. Conversely, flow from a high pressure to a low one is called a **favourable pressure gradient**. A favourable pressure gradient not only inhibits separation, but also slows down the rate of boundary layer growth, and delays transition from laminar to turbulent flow.

On road vehicles, the laminar boundary layer is normally confined to the front part of the body, where by sensible design it quite easy to maintain favourable pressure gradients. On vehicles prior to about 1980, the laminar layer was nearly always destroyed by the decorative grilles and sharp angles used on the front surface.

The process of large-scale mixing in a turbulent boundary layer is much more rapid than the mechanism of molecular impacts and mixing in a laminar layer, so a turbulent boundary layer will withstand a much more adverse pressure gradient than a laminar one. This is an important property that can be exploited, as shown later.

REATTACHMENT

Sometimes, a separated flow will reattach, as illustrated in Fig. 1.9. Between the points of separation and reattachment, a region of recirculating air known as a **separation bubble** is formed. In an open cabriolet the whole passenger compartment is immersed in a large separation bubble.

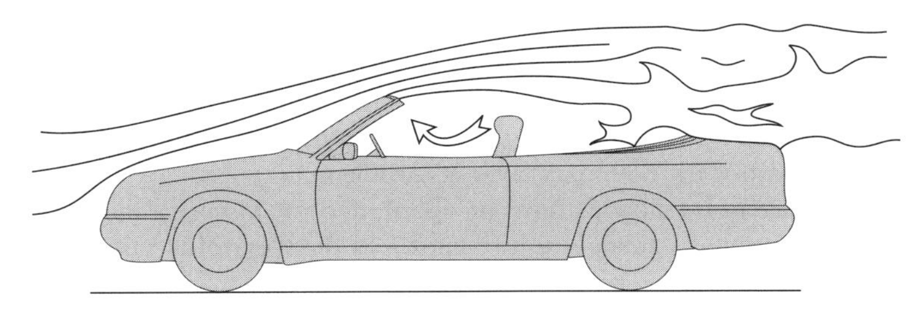

Fig. 1.9 Separation and reattachment
A large separation bubble forms in an open cabriolet.

REYNOLDS NUMBER

Since the type of boundary layer and its thickness influence the surface friction drag, and any flow separation, it is important to know what factors control the growth and form of the layer.

It has already been mentioned that the behaviour of the layer is strongly influenced by how the pressure varies along the direction of flow. Other important factors are the speed, and the density and viscosity (the stickiness) of the air; also, for a given geometric shape, the size of the vehicle is important.

To understand why size is important, consider model and full-scale versions of the shape shown in Fig. 1.10. Since transition to a turbulent flow is related to the

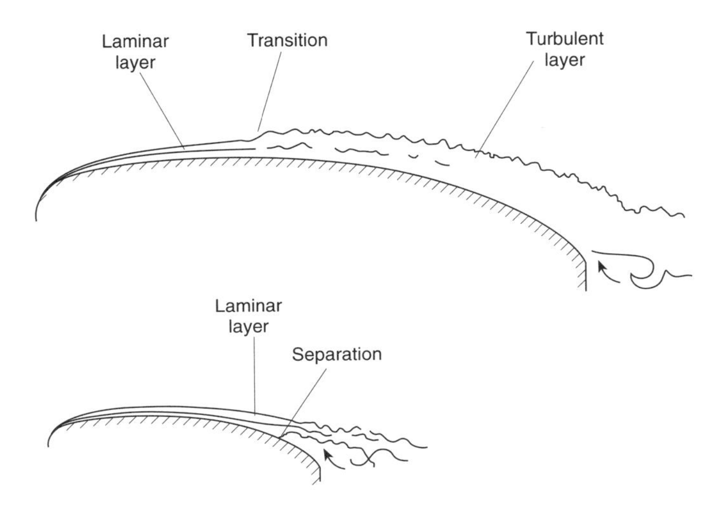

Fig. 1.10 Scale effect
Full-scale and model of the same shape showing differences in the flow field, caused by boundary layer effects. On the model the boundary layer is laminar over most of the length, and hence is likely to separate before reaching the rear. On the full-scale vehicle, the boundary layer has become turbulent near the front, and is better able to stay attached in the area of adverse pressure gradient at the rear.

distance from the front, it follows that at the same speed, there will be a much larger proportion of laminar flow on the model. More importantly, if as illustrated, the boundary layer on the model is still laminar after reaching the region of unfavourable pressure gradient, it will tend to separate earlier than on the full-scale vehicle, where most of the boundary layer is turbulent.

The dependence of the flow patterns on speed, density, viscosity and length can be expressed in terms of a single quantity known as the **Reynolds number**:

$$\text{Reynolds number} = \frac{\text{density} \times \text{speed} \times \text{length}}{\text{viscosity}}$$

or in mathematical symbols

$$\text{Re} = \frac{\rho v l}{\mu}$$

Reynolds number is just a number like a ratio, and the value will be the same regardless of whether the various quantities are measured in metric, Imperial or Federal units. For fixed values of air density and viscosity, the Reynolds number is effectively dependent on the vehicle size and speed only.

If the Reynolds number is increased by increasing the speed of the vehicle, the transition position moves forward, and the boundary layer becomes thinner. It can therefore be seen that the value of the Reynolds number is important in determining the type of flow around the vehicle.

TURBULENCE

Another important factor controlling boundary layer behaviour (and hence the nature of the flow around the vehicle) is the turbulence properties of the oncoming air. For a vehicle travelling along a road on a windless day, there will be virtually no air turbulence, but in a wind-tunnel, or on a windy day, the level of turbulence present can significantly modify the behaviour of the boundary layer. Increasing the free-stream turbulence tends to move the boundary layer transition position forwards, but the increased turbulence can also help to keep the flow attached.

There are many factors which characterize turbulence. The most important in terms of boundary layer development is the turbulence intensity of the streamwise velocity fluctuations, I_u, which is measured by the ratio of the standard deviation of the velocity fluctuations to the mean velocity.

$$I_u = \frac{\sqrt{\overline{u^2}}}{\overline{U}}$$

where u is the increase or decrease in velocity above the mean value $\overline{U}$. The overbar denotes a mean value.

The fluctuations in velocity along the mean flow direction are accompanied by unsteady sideways and vertical flow components v and w, so there are corresponding intensities I_v and I_w associated with these directions.

Turbulence is characterized by eddies, or swirling turbulent motions of differing sizes. The scale of these motions is indicated by the **integral length scale**, which is related to the average size of the eddies. The mathematical definition of integral length scale may be found in ESDU data sheet No. 74030.[3] The length scale of the turbulence influences the behaviour of the flow around a vehicle and the aerodynamic forces produced. There are many other characteristic measures of turbulence, but these need not concern us for the present.

THREE-DIMENSIONAL BOUNDARY LAYERS

Thus far, only the simple case of a boundary layer formed by air flowing directly from the front to the back of a vehicle has been considered. In reality, there are often strong sideways or *cross-stream* components of velocity on the surface of a road vehicle, and this considerably complicates the formation and behaviour of the boundary layer. Once again, it is necessary to remember that road vehicles and their air flow patterns are highly three-dimensional. It is important to avoid simply looking at the centre-line shape as if it extended to infinity on both sides. Curvature of the streamlines near the surface due to cross-stream components tends to provoke early transition of the boundary layer, reducing the extent of the laminar region. By contrast, cross-stream flows can sometimes help to keep the boundary layer attached; the sideways flow can vent a high-pressure region, thereby making the pressure gradient less adverse. The way in which cross-flows help to maintain attached flow over quite severely sloping rear screen surfaces on hatchbacks is described in Chapter 4.

VORTICES

There are often regions of flow which are dominated by vortices: the swirling flow structures which occur in nature as whirlwinds. Typically vortices may form around the A-posts (the framework on either side of the front windscreen) and at the rear, as shown in Fig. 1.11. Any vehicle that produces lift, either positive or negative, will leave a pair of vortices trailing behind it, and as described in later chapters, vortices represent a source of drag and also noise.

In the outer portion of a vortex, the speed reduces with distance from the centre (the radius).

$$\text{speed} = \frac{\text{a constant}}{\text{radius}}$$

The converse of this is that the speed increases towards the centre, and if the above law held right up to the centre, the speed would reach infinity there. In practice there is always a central **core** where the speed is high, but is no longer inversely proportional to the radius. In the core there is viscous slipping or shearing between adjacent layers of fluid.

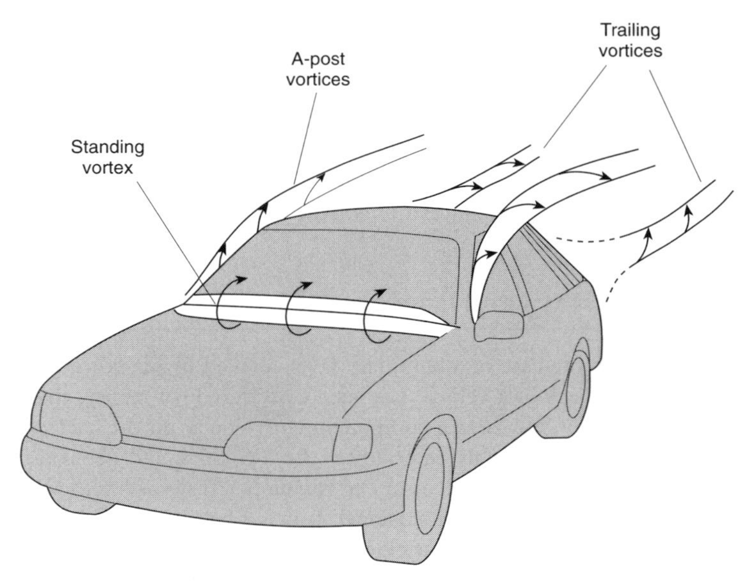

Fig. 1.11 Vortex formation
Vortices commonly form around the A-posts, and at the rear of a vehicle.

Because the speed is high in the centre of the vortex, the pressure and temperature are correspondingly low. The wings of racing cars generate very strong vortices at their tips, and on damp days these are made clearly visible due to the water vapour condensing in the low temperature (see Fig. 6.25). Vortices represent regions of low pressure, and because the rear surfaces of the vehicle are exposed to their influence, a rearward drag force will be produced.

An important property of vortices is that they cannot exist along a line of finite length in space. They must either terminate in a surface, or form a closed loop, as in a smoke ring. The vortex loops that form behind vehicles are dealt with in more detail in later chapters.

VORTEX STREETS

When bluff bodies are exposed to an air flow, a stream of alternating vortices can be formed, as illustrated in Fig. 1.12. This is known as a Karman vortex street. Because the vortices are shed alternately from either side of the body, the whole flow field alternates, and alternating aerodynamic forces are produced on the body. This mechanism is the source of the 'singing' of telephone wires in a high wind. It is also the source of the loud noise often generated by luggage-racks on vehicles.

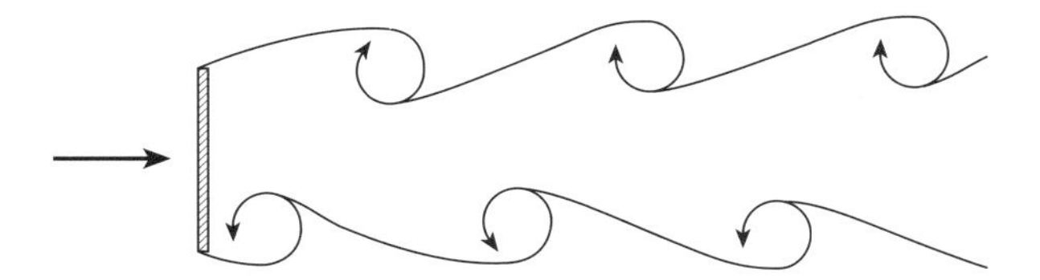

Fig. 1.12 A Karman vortex street
Vortices are shed alternately from the two edges of the body at a regular frequency which depends linearly on the flow speed and the inverse of the body depth.

The frequency of vortex shedding is called the Strouhal frequency, and it can be determined from the **Strouhal number** which is given by

$$\text{Strouhal number} = \frac{\text{frequency} \times \text{body depth}}{\text{flow speed}}$$

For a long-span circular cylinder, the Strouhal number is around 0.2. For a flat plate held broadside to the flow it is around 0.145.

The Strouhal number varies with the cross-stream *span* of the generating body, and can be as low as 0.09.[1] As the span of the body reduces, the shedding becomes unsteady, and eventually a well-defined street will not be found. A short-span object such as a wing-mirror will not usually generate a well-defined vortex street, although the turbulence it generates will tend to show fluctuations concentrated around the Strouhal frequency.

THE NATURAL ATMOSPHERIC WIND

It is easy to forget that the air flow over a road vehicle is also dependent on the local atmospheric wind. At ground level, the wind environment corresponds to a turbulent boundary layer with a thickness of about 100–500 m, depending on the local topology and meteorological conditions. Humans and their cars therefore inhabit the very bottom of this layer. For road vehicle engineering purposes, the variation of wind velocity with height can be represented accurately enough by a simple power law of the form

$$\frac{\overline{U}}{\overline{U}_{(\mathrm{ref})}} = \left(\frac{z}{z_{(\mathrm{ref})}}\right)^{\alpha}$$

where $\overline{U}$ is the mean velocity at height z, and $\overline{U}_{(\mathrm{ref})}$ is the mean velocity at a convenient reference height $z_{(\mathrm{ref})}$.

The reference height most commonly used is 10 m, as this is a standard height for mounting wind-speed measuring devices.

The index α depends on the roughness of the local terrain, and varies from about 0.15 for open country with no obstructions to 0.4 or more for city centres. Note that for a smooth flat plate, the value of α is found to be about 0.1428, or 1/7. This curve is plotted for the boundary layer profile shown in Fig. 1.6. Further details concerning the properties of the atmospheric wind may be found in the ESDU data sheets,[2–4] which provide a useful introduction to the topic.

The atmospheric wind produces a number of problems in terms of aerodynamic analysis and testing. Firstly, because the wind environment corresponds to a turbulent boundary layer, the intensity of the turbulence is higher than in a normal wind-tunnel. Equally as important, the natural wind contains eddies that are large in relation to the size of the vehicle, and it is virtually impossible to generate such large-scale turbulent structures in a wind-tunnel.

Another problem is that the wind will not in general be blowing directly on to the front of the vehicle, so there will be a cross-flow component of relative velocity, as shown in Fig. 1.13. This means that we need to take account of how the drag and other aerodynamic forces change with the relative wind direction. The angle between the lengthwise axis of the car and the resulting air flow direction is called the yaw angle.

In addition to the above effects, the relative flow direction varies with the height above the road. This comes about because the speed of the cross wind varies with height, and as can be seen from Fig. 1.14, the resultant velocity vector consequently changes direction with height. This effect is difficult to simulate correctly in a wind-tunnel, because if the model is simply mounted at a yawed angle to the flow, the flow direction will not vary with height. The only solution is to use a moving model running across the tunnel on a track, as described in Chapter 11.

The most difficult aspect to cope with when dealing with the atmospheric wind is that it is both unsteady and unpredictable, which means that we have to resort to using statistical probabilities rather than definite values. For this reason, except when dealing with stability analysis and performance calculations, vehicle aerodynamicists tend to ignore the effects of the natural wind, and design primarily for still air conditions.

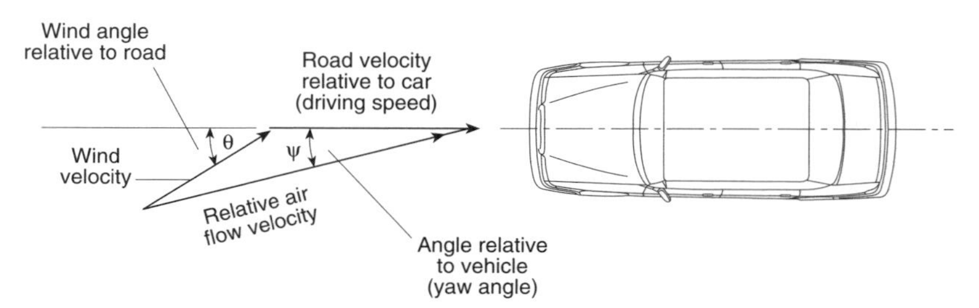

Fig. 1.13 Components of relative air velocity in a cross-wind.

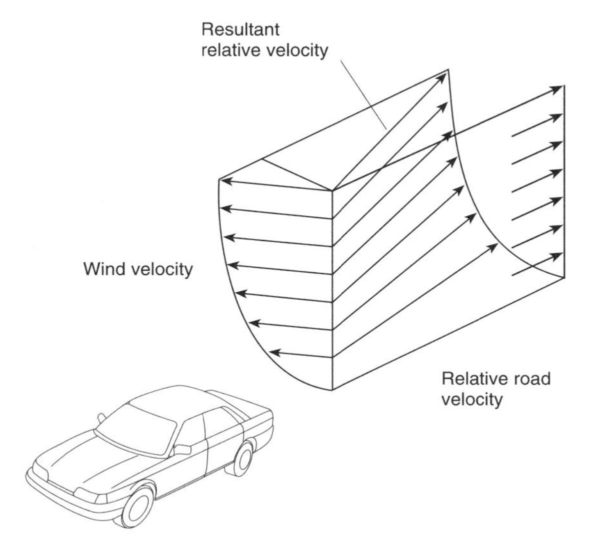

Fig. 1.14 Skewing of the relative wind velocity vector with height.
(*After Howell,*[5] *courtesy of J. Howell, Rover Group Ltd*)

REFERENCES

1. Barnard R. H., Vortex shedding under realistic conditions, *Proc. Symposium: Designing with the Wind*, CSTB Nantes, paper VII-4, 1981.
2. ESDU, data sheet No. 72026, Characteristics of wind speed in the lower layers of the atmosphere near the ground, Engineering Sciences Data Unit, London.
3. ESDU, data sheet No. 74030, Characteristics of atmospheric turbulence near the ground, Part 1: definitions and general information, Engineering Sciences Data Unit, London.
4. ESDU, data sheet No. 74031, Characteristics of atmospheric turbulence near the ground, Part 2: single point data for strong winds (neutral atmosphere), Engineering Sciences Data Unit, London.
5. Howell J. P., The side load distribution on a Rover 800 saloon car under crosswind conditions, *Proc. 2nd Wind Engineering Society Conference*, Warwick, 1994.

CHAPTER 2

DRAG AND LIFT FORCES

AERODYNAMIC DRAG

For the design of normal road-going vehicles, the most important aerodynamic factor is the drag force. The total force resisting the forward motion of a road vehicle comes partly from the rolling resistance of the wheels, and partly from aerodynamic drag. As will be shown in the next chapter, the aerodynamic drag dominates at speeds above about 65–80 km/h (40–50 mph), and there are considerable economic and performance advantages to be gained from drag reduction. In this chapter the various sources of aerodynamic drag are described, and some basic design principles for drag minimization are given.

THE IMPORTANCE OF AERODYNAMIC LIFT

Road vehicle aerodynamic research was initially concentrated on the problem of reducing drag, but it soon became apparent that lift forces were also of some importance. Lift indirectly affects the drag, as will be shown later, but more significantly, considerable improvements in the roadholding and stability can be obtained by reducing lift, or even generating a down force (negative lift); the roadholding improvement is of particular importance in the case of racing cars. Therefore, the salient features of lift production on road vehicles will also be described in this chapter.

DRAG COEFFICIENT

It is useful to have some simple means of comparing the aerodynamic drag produced by different vehicle shapes regardless of their size or the driving speed. This is conveniently provided by a factor called the **drag coefficient** (C_D) which is mainly dependent on the shape of the vehicle. In addition to this shape-related coefficient, the aerodynamic drag also depends on the frontal area of the vehicle, the air density, and the square of the relative air speed. The relationship between drag and these factors can be expressed by

$$\text{Drag} = \tfrac{1}{2}\rho V^2 A C_D \qquad [2.1]$$

where A is the frontal area
ρ is the density of the air
V is the speed of the vehicle relative to the air

The frontal area referred to above is the projected frontal area, which is illustrated in Fig. 2.1. For modern cars the projected frontal area is nearly always around 80% of the product of overall height and width.

It will be seen that the dynamic pressure ($\tfrac{1}{2}\rho V^2$) described in the previous chapter occurs in the above expression, and we could simply write

$$\text{Drag} = \text{Dynamic pressure} \times (\text{Projected frontal area}) \times C_D \qquad [2.2]$$

As stated earlier, the drag coefficient depends mainly on the shape of the vehicle, and from the relationships above, it can be seen that for cars of similar frontal area travelling at the same speed, the vehicle with the highest drag coefficient will produce the most drag. The drag coefficient thus provides a good indication of the relative merits of different vehicle shapes.

Knowing the drag coefficient and the projected frontal area, expression [2.1] can be used to work out the drag at any speed. Unfortunately, although the drag coefficient mainly depends on the shape of the vehicle, it is also dependent on a number of other factors such as the turbulence properties of the flow and the Reynolds number $(\rho V l)/\mu$. (Reynolds number was defined in the previous chapter.) The dependence on Reynolds number means that drag coefficient varies with speed, but fortunately, over the normal range of road vehicle cruising speeds, the changes are usually insignificant, and C_D can be treated as if it were a constant. It is only when testing small models in a wind tunnel that the variation of C_D with Reynolds number becomes important.

A useful feature of C_D is that it is a quantity with no units or dimensions (like a ratio), which means that it has the same value regardless of whether metric or Federal units are used.

Although the drag coefficient is denoted by C_D in most British and American sources, the alternative C_X indicating a drag coefficient in the longitudinal x-axis

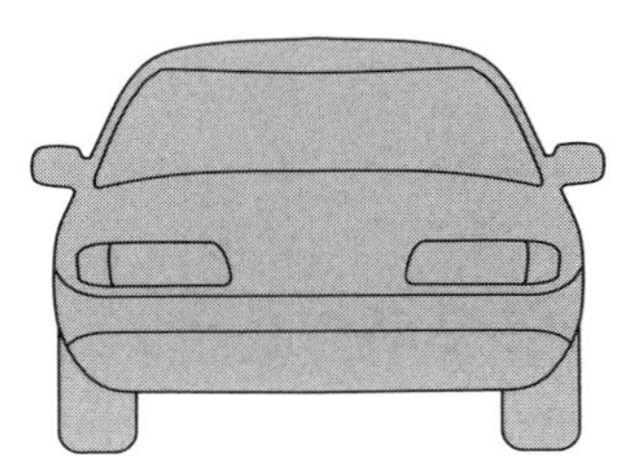

Fig. 2.1 Projected frontal area.

direction is often used, particularly in French and southern European sources. The German version C_W (Widerstands) is also frequently found in published research papers.

For an aircraft, a major contribution to the overall drag comes from the wing, and it is convenient to relate the drag coefficient to the wing plan area rather than to the projected frontal area. For this reason C_D values for cars are always numerically much larger than those of aircraft.

DRAG OR SHAPE FACTOR

It can be seen from expressions [2.1] and [2.2] above that the drag depends on the product of the frontal area and the drag coefficient, and not just on the drag coefficient. A low drag coefficient does not necessarily mean low drag. For example, the VW Microbus of 1958 (Fig. 2.2) surprisingly had a lower drag coefficient than that of its contemporary, the D-type racing Jaguar (Fig. 2.3). The Microbus, however, produced far more drag because of its larger frontal area.

The product of C_D and area (A) is sometimes called the drag factor. Table 2.1 shows some interesting comparisons between different vehicles using both drag factor and drag coefficient. The table includes a number of historical vehicles, and is based on data published in a variety of sources. It has become common for manufacturers to use low C_D values as a selling point, but Table 2.1 shows that drag coefficient alone can be a misleading indicator.

Fig. 2.2 The VW Devon Caravette, (based on the VW Microbus), with a lower than expected C_D of 0.45.
(*Photograph courtesy of R. Davies*)

Table 2.1 C_D and C_DA values for various vehicles.

Vehicle	C_D	C_DA
Bugatti type 51 (1933)	0.74	0.96
VW Beetle	0.48	0.87
Jaguar D-type (1955)	0.49	0.59
Citroen DS (1958)	0.37	0.81
VW Microbus (1958)	0.45	1.04
Ford Sierra	0.34	0.67
Tiga G83 Group C Sports racer (1983)	0.24	0.38
Formula 1 G.P. Slow circuit	1.00	0.98
GM Calibra	0.26	0.49

Fig. 2.3 The D-type Jaguar with a higher than expected C_D of 0.49.

CONTRIBUTIONS TO AERODYNAMIC DRAG

Aerodynamic drag is a force produced partly as a result of the distribution of pressure around the vehicle and partly by the frictional or shearing action of the flow over the surface. Figure 2.4 illustrates the shear and pressure forces on a small element of the surface. Adding together the lengthwise components of these two forces gives the drag on the surface element. Adding together the lengthwise components of the forces on all the elements of the whole surface gives the total drag.

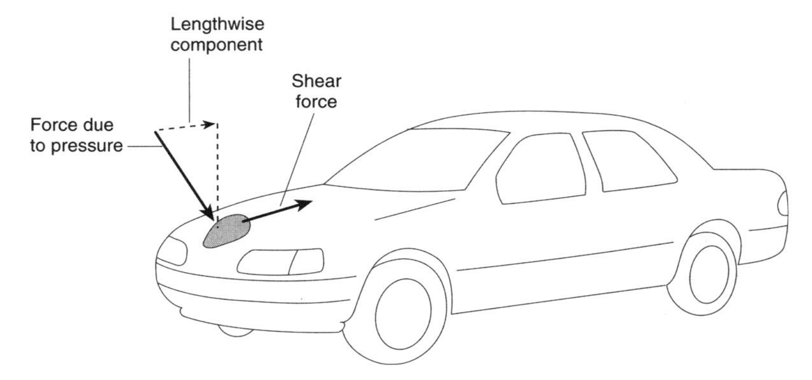

Fig. 2.4 Shear and pressure forces on a vehicle.

SURFACE FRICTION DRAG

For conventional road vehicles, the drag due to frictional shear is relatively small, but can be in the order of 30 per cent of the overall drag for a typical modern car. Carr[2] has estimated that the contribution of surface friction drag to the overall C_D values is in the order of 0.04 for a car with reasonably smooth profile, rising to 0.11 if the underside is rough (which it usually is on domestic cars). Carr's values were deduced from estimates of the surface friction coefficient for flat plates of equivalent surface area, and therefore give only a very approximate guide. The value of 0.11 is probably too high for recent car designs as some attention is now paid to underbody aerodynamic design.

PRESSURE OR FORM DRAG

The greater part of the aerodynamic drag on road vehicles comes from the fact that the pressure over rearward-facing parts of the surface is on average lower than on forward-facing ones.

The pressure distribution around a vehicle is produced by a number of interacting influences, one of the most important being the boundary layer. To appreciate how the boundary layer influences the pressure distribution, consider the simple case of two-dimensional flow around a smooth symmetrical shape, as illustrated in Fig. 2.5. Right on the nose, the relative air speed is brought to zero. The flow then accelerates, reaching a high relative speed in the vicinity of the widest part, and then slows down as it approaches the tail. From the Bernoulli relationship, given in Chapter 1, it can be seen that this means that there is a high pressure near the tip of the nose where the speed is low, and a relatively low pressure in the region of the widest part. Without the influence of viscosity and with no sharp corners, the streamlines would close up neatly behind the shape

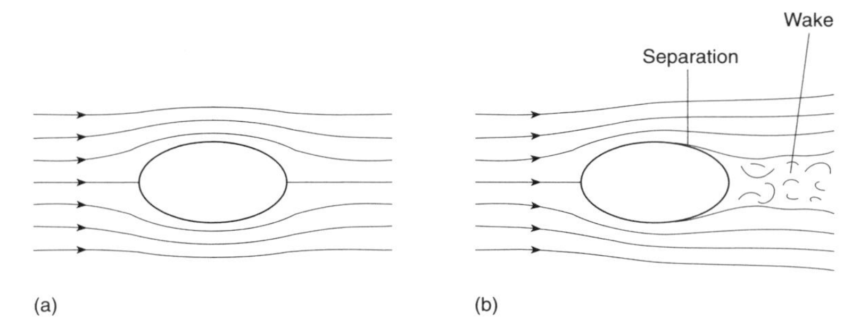

Fig. 2.5 The effects of viscosity
(a) Theoretical flow pattern obtained when the effects of viscosity are ignored; (b) typical actual patterns for a real air flow.

shown in Fig. 2.5, producing a symmetrical pattern as shown in Fig. 2.5(a). The pressure distribution would also be symmetrical, and would therefore produce equal and opposite forces on corresponding forward- and rearward-facing parts of the surface; there would thus be no drag. This result is predicted by the classical incompressible 'potential flow' theory, which assumes an imaginary 'inviscid' fluid: a fluid with no viscosity. (Potential flow theory may be found in most fundamental texts on fluid mechanics or aerodynamics: Houghton and Carpenter,[7] for example). In fact, the potential flow theory predicts that the pressure distribution would result in no net drag force on any smooth shape, whether symmetrical or not.

In reality, because of the effects of viscosity, the streamlines around the shape shown would look more as in Fig. 2.5(b). This is because, as has previously been described, viscosity causes the energy in the air flow close to the surface (the boundary layer) to degrade to forms from which the process cannot be passively reversed; the energy is for practical purposes lost. As a consequence of the energy degradation, the air cannot return to its initial or upstream values of speed and pressure; instead, a 'wake' of slower moving air is formed. The boundary layer may even separate (fail to follow the contours of the shape), as depicted in Fig. 2.5(b). In this real flow, the streamline pattern and pressure distribution are no longer symmetrical. The pressure over the rear portion of the shape is on average lower than over the front, and there will therefore be a net rearward drag force. Flow separation was described in Chapter 1.

The contribution to overall drag arising in this way is commonly known as **form drag**, but the more modern term **boundary layer normal pressure drag** is now preferred, at least in academic circles. The newer term may be a bit long-winded, but it does conveniently identify this type of drag as being associated with the pressure distribution and boundary layer effects. For convenience, it is often shortened to **pressure drag**. At one time, this type of drag represented the major contribution, but on modern designs it accounts for only about one third.

When flow separation takes place, the amount of boundary layer normal pressure drag produced depends largely on where flow separation occurs. A circular plate held normal to the flow will produce separation around the periphery of the leading face resulting in a wide wake with a high drag coefficient. At the other extreme, a teardrop shape with a long tail can retain attached flow right to the end, with a consequently low coefficient. Table 2.2 shows some of the drag coefficient results for simple shapes quoted by Hoerner.[6] Note that the sharply pointed shape does not give the lowest drag, and that the apparently blunt hemispherical shape is much better. The pointed shape tends to produce flow separation at the base of the cone. Many people find this surprising result hard to believe, since it would seem logical that the pointed shape would be better able to penetrate through the air. This misconcept however is due to our habit of assuming that fluids behave in a similar manner to solid objects. The 'streamlined' coffin-shaped fairings that can be attached to roof-racks are invariably mounted pointed-end forwards, and have a blunt rear end. In reality they would produce less drag if mounted the other way around.

Table 2.2 C_D values for simple three-dimensional shapes. Note that for some of these shapes, the drag coefficient varies significantly with Reynolds number. The values are typical for Reynolds numbers in the range 10^4–10^6.

Shape		C_D
→	Sphere	0.47
	Hemisphere	0.42
	60° cone	0.5
	Circular plate	1.17
	Cube	1.05
	Teardrop [t/c = 0.25]	0.05

(Based on data from various sources presented in Hoerner[6])

SENSITIVITY OF DRAG COEFFICIENT TO SPEED AND SCALE (REYNOLDS NUMBER)

Reynolds number ($\rho Vl/\mu$) was defined in Chapter 1, and effectively combines the effects of speed (V) and scale or size (l). For shapes such as a circular rod, the position of flow separation depends strongly on the Reynolds number. At low Reynolds numbers (low speeds and/or small sizes) the boundary layer is smoothly laminar, and will separate early (as described in Chapter 1), producing a high drag coefficient. At higher Reynolds numbers, the boundary layer becomes turbulent before separation, and this delays separation. Moving the separation position aft produces a narrower wake with a corresponding drop in drag coefficient. The drop is large, as may be seen from Fig. 2.6.

Very streamlined shapes such as the teardrop shown in Table 2.2 are less sensitive to Reynolds number change because the flow does not separate. Bluff shapes such as the disc shown in Table 2.2 are also largely insensitive to Reynolds number, as the separation position is fixed at the sharp edges. Unfortunately modern cars, which may be described as being semi-streamlined, are usually sensitive to Reynolds number variation, because, as with the circular rod, the separation position is not fixed by sharp corners.

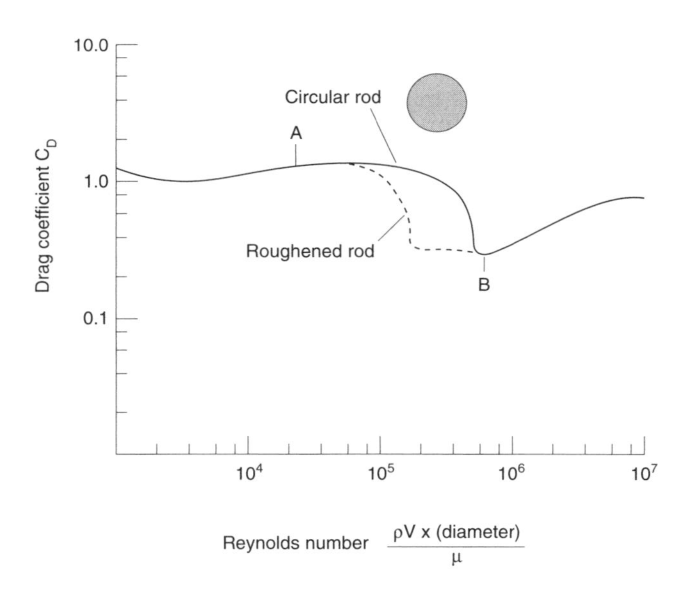

Fig. 2.6 The variation of C_D with Reynolds number for a circular cylinder; note the sudden and significant drop that occurs.

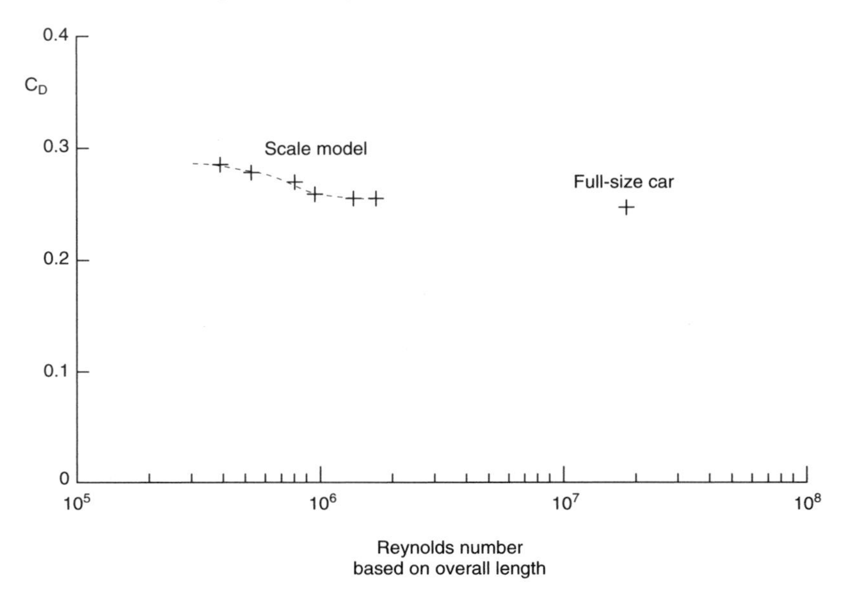

Fig. 2.7 Variation of C_D with Reynolds number for a Group C racing sports car tested by the author.

Figure 2.7 shows the variation of C_D with Reynolds number for a scale model of a racing sports car. A data point from a test on the full-scale vehicle with a relative wind speed of 113 km/h (70 mph) is also included. The highest Reynolds number for the model tests is far below the full-scale value; note that the graph has a logarithmic scale. The curve for the model data can be seen to be flattening towards a constant value at high Reynolds numbers, which in this case is close to the full-scale value, but for a different vehicle, it is possible that a sudden change might occur (as for the circular rod) beyond the limit of the measured model data. Fortunately, at Reynolds numbers above a value corresponding to about 40 m/s with a 1/4 scale model car, little variation in C_D is usually found. For a circular cylinder with a diameter equal to the height of a 1/4 scale model of a typical car, the above conditions correspond to a Reynolds number of around 9.6×10^5, which lies above the critical region shown in Fig. 2.6.

TRAILING VORTEX DRAG

In the explanation of pressure drag above, a two-dimensional flow has been considered: always a dangerous approach in automobile aerodynamics, because in reality, the flows are highly three-dimensional. Many of the approximations to two-dimensional flow that are commonly used in aircraft aerodynamic analysis cannot be used for road vehicles. Figure 2.8 shows something of the three-dimensional nature of the flow around a typical car. Most of such vehicles produce a lower pressure on the top surfaces than on the underside, and an important consequence of this is that they generate lift, as will be described later. An equally important consequence is that the air tends to flow from the high pressure underside, up the sides, and towards the top surfaces, resulting in the production of vortices in the wake, as illustrated in Fig. 2.8. These vortices, called **trailing vortices**, are similar in nature to whirlwinds. A large amount of energy goes into the formation of these large swirling masses of air, and it is clear therefore that they represent a source of drag. In the case of fastbacks, strong stable trailing vortices can form over the rear screen leading to very high drag, as described in Chapter 4.

Some understanding of the mechanism by which the trailing vortices generate drag can be gained by reference to Fig. 2.9. This figure compares the centre-line flow patterns of a theoretical two-dimensional flow (with no trailing vortices) with the realistic three-dimensional case. In the three-dimensional case depicted, the vortices draw air away from the rear of the vehicle, creating a low pressure there, and thereby pulling the flow down. Because the air is swirling with a high speed, the pressure in the vortex is low, as predicted by the Bernoulli relationship, and therefore any surface exposed to the influence of a vortex will be subjected to a reduced pressure. A reduced pressure over the rear of the vehicle will obviously increase the drag. It is also possible to deduce from momentum considerations that if the air is being pulled towards the rear of the car, the corresponding reaction will pull the vehicle backwards.

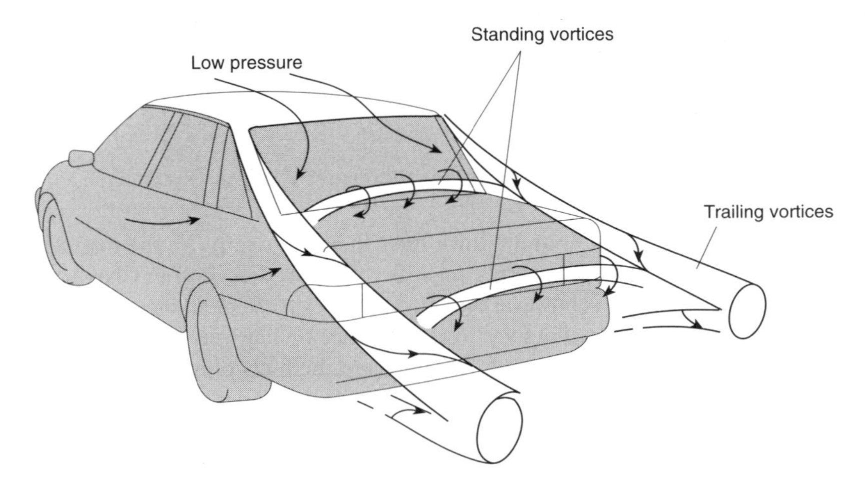

Fig. 2.8 The three-dimensional nature of the flow around a road car, and the origins of trailing vortex drag.

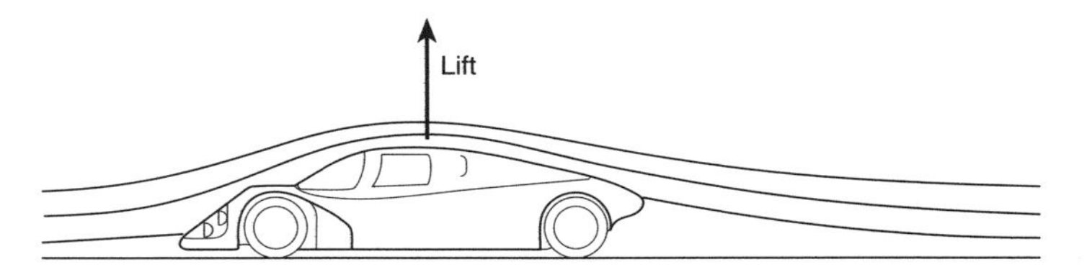

(a) Flow patterns without the effect of trailing vortices

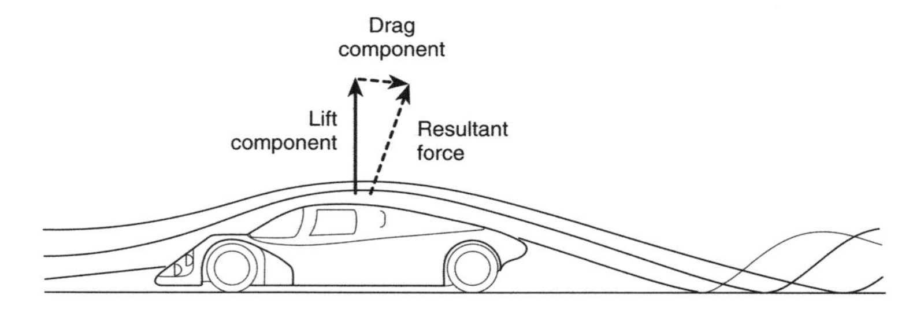

(b) Real flow patterns with the effect of trailing vortices

Fig. 2.9 Centre-line flow pattern for: (a) two-dimensional flow, showing flow patterns without the effect of trailing vortices; (b) three-dimensional flow, showing real flow patterns with trailing vortices.

From the above, it may be seen that the trailing vortices produce drag by modifying the pressure distribution around the vehicle. This contribution to drag is commonly called **trailing vortex drag**, although the older aeronautical term 'induced drag' is still sometimes used.

TRAILING VORTEX PRODUCTION BY RACING CAR WINGS

The use of inverted wings to produce down force in order to improve the roadholding of racing cars has been standard practice for many years, as described in Chapter 6. The improved roadholding is however obtained at the expense of an increase in drag, because wings also generate trailing vortices, and hence trailing vortex drag.

As shown in Fig. 2.10, the air tends to flow from the high-pressure surface to the lower pressure surface by spilling around the ends of the wings. The trailing vortices therefore emanate from the wing tips. As on the vehicle body, the vortices bend the oncoming flow so that the resultant force vector is tilted back, producing a rearward drag component. Barnard and Philpott[1] describe the generation of trailing vortex drag by a wing in more detail.

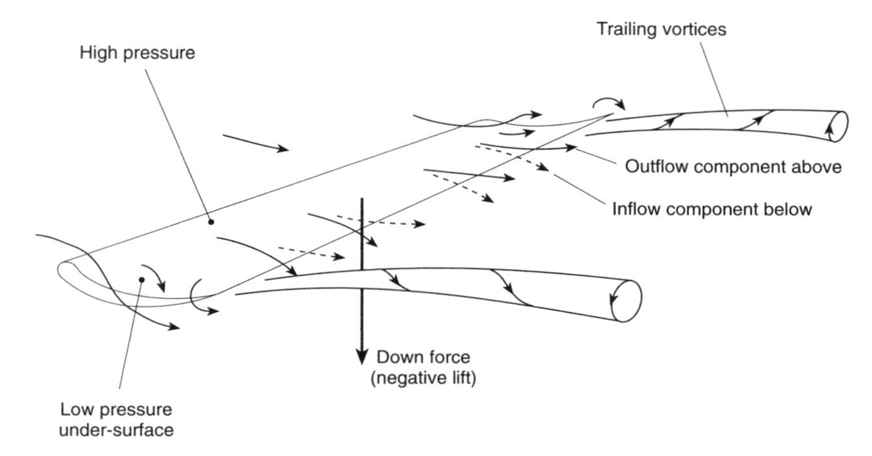

Fig. 2.10 Trailing vortex production by the inverted wing of a racing car. Air tends to flow outward from the high pressure on the upper surface, and inward to the low pressure on the under-surface, curling round the tips. Trailing vortices form at the tips and there is considerable vorticity in the wake.

THE GENERATION OF LIFT BY A ROAD VEHICLE

From the above, it may be seen that trailing vortex drag production is associated with the differences in pressure between upper and lower surfaces, and hence with lift; it will therefore be helpful to describe lift generation at this stage. Apart from its association with trailing vortex drag, lift is an important factor in road vehicle aerodynamics in its own right, having a strong influence on stability, roadholding and performance, as described in subsequent chapters. In the case of racing cars, the production of negative lift or down force is the major aerodynamic preoccupation.

For an object flying through the air away from the influence of the ground, two factors that affect the amount of lift force generated are the **angle of attack**, and the **camber** or curvature of the body. As illustrated in Fig. 2.11, increasing either will increase the lift. For road vehicles, however, the proximity of the road has a profound effect. If the car were in intimate sliding contact with the road, as in Fig. 2.12(a), the air flow over the top would have to accelerate, and lift would result from the low pressure on the upper surface. In reality, however, some air always flows underneath, and the lift is strongly dependent on the underbody flow and pressure distribution. Sealing the rear of the underbody as in Fig. 2.12(b), and the sides, would expose the underside to the high stagnation pressure at the front, and would therefore produce positive lift. By contrast, a combination of an air dam or seal at the front and a diffuser at the rear as in Fig. 2.12(c) ensures that the underbody is at a low pressure, which therefore tends to generate negative lift or down force. This latter method is used in racing cars, as described in more detail in Chapter 6.

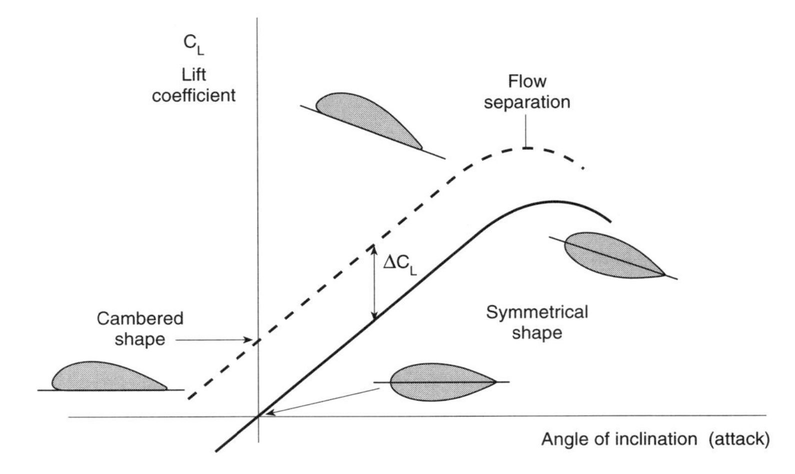

Fig. 2.11 For a free-flying object, lift is produced by inclining a shape to the flow and/or by giving it a cambered (curved) form. The cambered shape gives more lift at any given angle of attack.

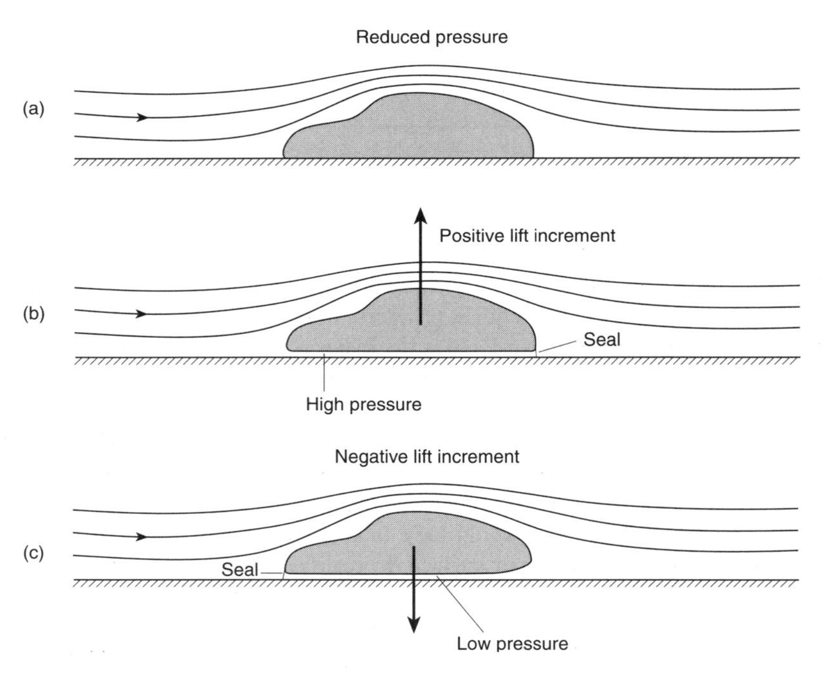

Fig. 2.12 A vehicle in intimate sliding contact with the road would generate lift due to the low pressure on the upper surface associated with the high air flow speed. When there is a clearance gap, however, the lift can be positive or negative depending on whether the undersurface is vented to the high pressure at the front or the low pressure at the back: (a) sliding contact; (b) underside vented to high pressure at the front; (c) underside vented to low pressure at the rear.

Normal road-going vehicles fall somewhere between the extremes of a free-flying object and one dominated by ground effect, which makes generalizations about lift characteristics difficult. Camber and angle of attack are important, but these influences are considerably distorted by the proximity of the ground.

Lift is produced when the average pressure on the top of an object is lower than that below. The detailed mechanism by which this difference comes about is not easy to describe in simple physical terms, and there have unfortunately been many misguided and misleading attempts. It is important to note that no special geometry is required, and almost any shape, even a brick, will generate lift if inclined to the flow direction. Barnard and Philpott[1] give some insight into the generation of lift by a wing.

LIFT COEFFICIENT C_L

Lift coefficient is defined in a similar way to drag coefficient:

Lift = Dynamic pressure × Frontal area × C_L

Note that for convenience, the reference area is still the **projected frontal area**, even though the lift is more directly related to the plan area. In aeronautical engineering, the wing area is used as the reference area, and consequently C_L data for wing sections are also normally based on their plan area, not their frontal area.

Common alternative symbols for C_L are C_Z, indicating the *Z*-axis, and the German C_A indicating Auftriebsverhalten (lift).

If the flow is attached, as on an aerofoil at small angles of attack, the lift increases linearly with angle of attack, but once the flow starts to separate, the lift begins to fall off. For a wing or aerofoil, this separation or **stalling** occurs at around 12° to 15° angle of attack, but for a car shape, the complex geometry and high degree of three-dimensionality mean that the stalling angle is not readily predictable, and may be as small as two or three degrees. Note that below stall, the influences of camber and angle of attack are largely independent; the rate at which the lift increases with angle of attack is not affected by the degree of camber, and increasing the camber produces the same increment of lift at any angle of attack.

In the 1970s when serious interest in vehicle aerodynamics was aroused, it was found that cars of that time gave lift coefficients that were typically in the range 0.3 to 0.5: roughly similar in magnitude to their drag coefficients. With attention to aerodynamic design C_L values of road-going cars have been reduced considerably, and in recent good designs C_L is nearly zero. Racing cars are designed to give a down force rather than lift, and therefore have negative lift coefficients.

THE CONNECTION BETWEEN C_D AND C_L

For aircraft it is found that under normal cruising conditions, the contribution to drag coefficient from trailing vortices is proportional to the square of the lift coefficient.

C_D is proportional to ${C_L}^2$

It is not always appreciated that the relationship above holds for aircraft only when the flow is fully attached (as in normal cruising flight). It is not valid for separated flows, and it cannot therefore be applied with confidence to road vehicles. Another difference between aircraft and ground vehicles is that for aircraft wings, the trailing vortices have little effect on the wing boundary layer, and thus have only a minor influence on the boundary layer normal pressure drag or the surface friction drag; the mechanisms of pressure drag, surface friction drag and trailing vortex drag are largely independent. Unfortunately, for road vehicles this is not the case. Unlike cruising aircraft, cars almost invariably produce significant regions of separated flow. Changes in the strength of the trailing vortices can alter the positions at which the separation occurs, and thus can profoundly affect the pressure drag.

The differences of approach between road vehicle and aircraft aerodynamics have been mentioned a number of times. This is because aircraft aerodynamic theory has been evolved over many years, and it is all too easy to fall into the trap of attempting to apply the well-developed aeronautical principles to cars, forgetting that a number of assumptions are involved in the aeronautical theory that may not be valid for automotive applications.

Despite the interaction between the trailing vortices and the pressure drag mechanism in car flows, it is found that there there is a general trend to follow a law of the form

$$C_D = C_{D_0} + kC_L^2 \qquad [2.3]$$

where C_{D_0} represents the combined contribution due to the pressure and surface friction drag, and k is a constant. Figure 2.13, for example, shows C_D plotted against C_L^2 for a wind-tunnel model of a sports car with various spoiler and shape

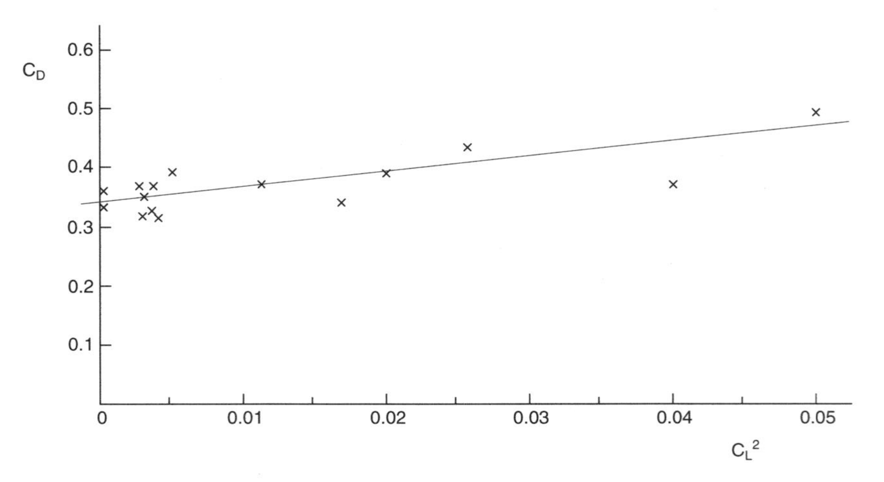

Fig. 2.13 The connection between lift and drag coefficient for a model sports car with various spoiler and styling modifications tested in the wind-tunnel at the University of Hertfordshire.

modifications. There are some obvious anomalies, and a great deal of scatter, but generally the data show a trend of the form given above.

Carr[2] examined the relationship between C_L and C_D for a large number of vehicles, and deduced a value of approximately 0.5 for the constant k in equation [2.3] above. The scatter of the data was also found to be wide, and all that can be safely deduced is that, in general, increasing the lift will increase the drag. For racing car wings, however, the common aeronautical expression can be used, namely

$$C_D = C_{D_0} + k_1 C_L{}^2/\pi AR$$

where C_{D_0} is the drag coefficient for all except the trailing vortex drag
k_1 is a constant mainly dependent on the planform shape
AR is the aspect ratio (wing span/wing chord)

The proximity of the trailing vortices to the ground affects the amount of drag produced; if the vortices are formed close to the ground, they are weakened. The effect is described in more detail later in this chapter in the context of drag reduction. The important connection between the lift and drag of racing cars is discussed in some detail in Chapter 6.

EXCRESCENCE DRAG

Any protrusions from the surface or gaps between panels will produce drag. Part of the cause is the small areas of low pressure that are produced by locally separated and often vortical flows. The disturbance caused by such excrescences may also produce drag by provoking transition of the boundary layer, triggering separation of the main flow, or producing a thickening of the boundary layer. Cresswell and Herz[4] estimate that driving mirrors on large trucks can increase the overall drag by 10 per cent.

THE INFLUENCE OF YAW ANGLE

Thus far, only the case of a vehicle driving through still air has been considered. In general, there will be some atmospheric wind blowing, and this will meet the vehicle at an angle. As described in the previous chapter, the angle between the vehicle longitudinal axis and the resultant relative wind direction is called the yaw angle (ψ). It is found that the drag coefficient changes significantly with yaw angle, the rate of change being dependent on the type of vehicle, and its length to width ratio. Figure 2.14 shows the variation of C_D with yaw angle for various types of vehicle. It will be seen that for large lorries the dependence of drag on yaw angle is strong. In the next chapter, a description is given of the procedure for obtaining a **wind-averaged** C_D value which takes account of variations in the atmospheric wind speed and direction.

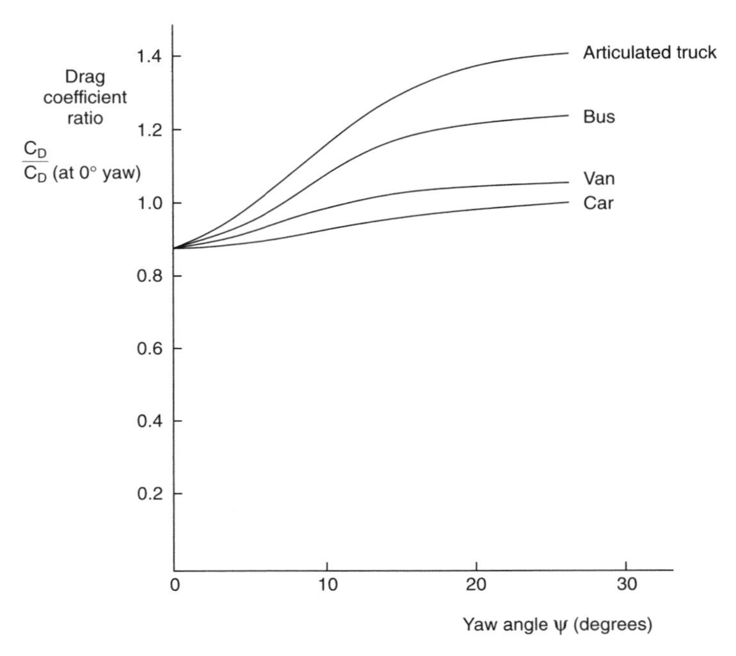

Fig. 2.14 The form of the variation of C_D with yaw angle for various types of vehicle. (*Based on data from a number of sources*)

DRAG DUE TO WHEELS

There have been a several studies of wheel drag and lift, but the agreement between results is not particularly good. This is partly because many of the early wind-tunnel tests such as those by Scibor-Rylski[11] were conducted at low Reynolds numbers (small-scale models at low speeds), and also because the data are sensitive to quite small differences in tyre geometry. Lift coefficient values also suffer from the fact that it is very difficult to measure the lift on a rotating wheel in contact with the road, as explained in Chapter 11.

Exposed wheels produce high drag coefficients due to the large highly turbulent wake, and the presence of strong trailing vortices. Figure 2.15 shows the centre-line flow and pressure distribution for an exposed rotating wheel. The top of the wheel is moving into the air stream at twice the driving speed, and is trying to drag the air forward against the stream. This has the effect of causing the flow to separate very early, and as shown, the separation position is usually in the front top quadrant. The consequence is a broad wake, and high drag. The side view of course only tells a small part of the story; the tyre width is always less than the diameter, so the flow around the sides is equally important. The side flow has complicated structure, and contains strong vortices which trail downstream, giving high associated drag.

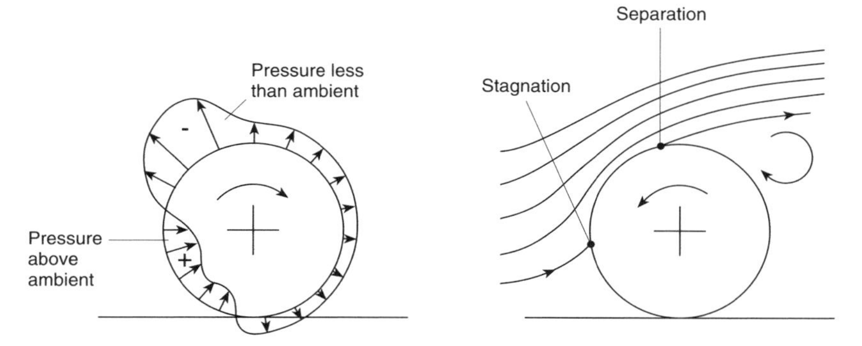

Fig. 2.15 Centre-line pressure distribution and streamline patterns for a rotating wheel.

Exposed tyres also generate lift, since all the air has to go over or around the tyre; contact with the road ensures that it cannot go underneath. By contrast, a rotating cylinder in a free stream of air (away from the influence of the road) will generate a downward or negative lift force by what is known as the Magnus effect. The direction of rotation of the cylinder determines whether the lift is up or down, and in the case of car wheels, the directional sense would generate a down force, but the presence of the road effectively kills the Magnus effect, and as stated, tyres in contact with the road always produce positive (upward) lift.

Figure 2.16 shows drag coefficient values (based on the frontal area of a single

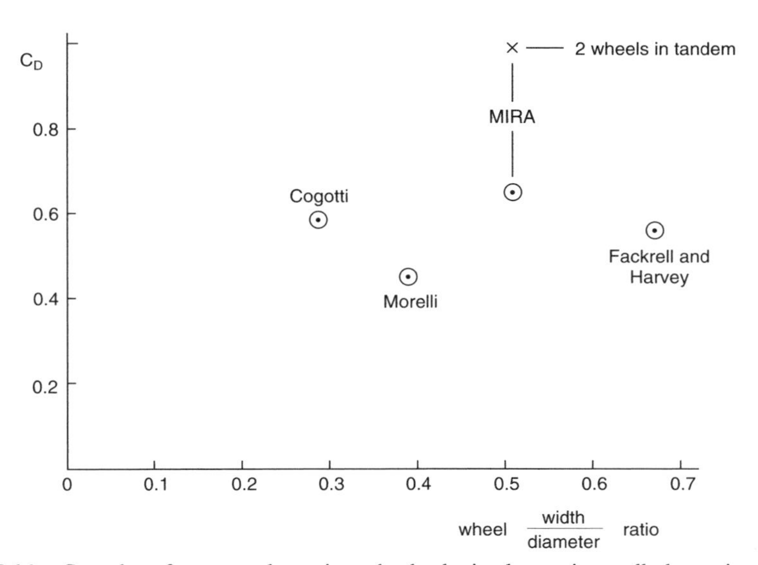

Fig. 2.16 C_D values for exposed rotating wheels obtained experimentally by various researchers (see references for further details).

tyre) obtained by various researchers. These are for isolated unfaired rotating wheels in contact with the road. The values are plotted against tyre width/diameter ratio. All but the lowest ratio represent wide racing tyres. The most useful result is probably that obtained by Stapleford and Carr[12] who measured the lift and drag on wheels in tandem, that is, one behind the other, corresponding to the combined effect of front and rear wheels. The front wheel partially shields the rear wheel, and therefore, as may be seen, the total drag on two wheels in tandem is less than twice the value for one isolated wheel. For a single isolated wheel, the drag coefficient is on average around 0.6, but is about 1 for the combined drag of two wheels in tandem (based on the projected frontal area of one tyre).

There is far less agreement between lift coefficient data from the various researchers, but this is due to the difficulty of measuring lift force on a wheel which is in contact with the road, as explained in Chapter 11. The negative lift value obtained by Cogotti[3] is due to the presence of a small gap between the road and the wheel. The tread on road-going tyres allows some air to flow under the tyre, and hence tread pattern and wear can affect the lift.

ENCLOSED OR FAIRED WHEELS

Unfaired or exposed wheels are only used on some classes of racing car (and off-road plant), and the above data are therefore of limited direct use. On most road vehicles, the wheels are contained within the bodywork or within some form of fairing. Experimental data in Scibor-Rylski[11] and Morelli[9] suggest that the drag coefficient of faired wheels remains fairly constant at the exposed wheel value, as long as the coefficient is based on the *exposed* frontal area of the wheel (see Fig. 2.17), except at very small exposures, where the value tends to rise.

Using Stapleford and Carr's values for tandem wheels, and assuming that fairing the wheels does not change the C_D value based on exposed frontal area, the contribution to the overall drag coefficient of a car due to the presence of the

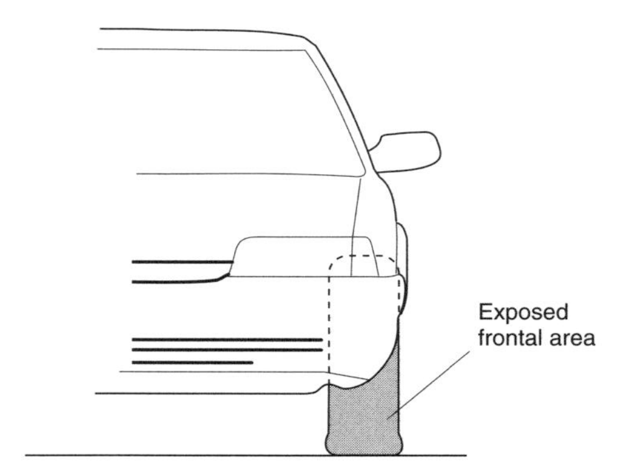

Fig. 2.17 Exposed frontal area in a faired wheel.

wheels may be estimated. Taking typical tyre and frontal area data for a medium-sized domestic car gives an estimated contribution to C_D from the wheels of around 0.05. Cogotti,[3] however, found that adding wheels to a highly low-drag body shape produced an increase of 0.084, despite careful attention to design detail intended to minimize drag. Similarly, Mercker *et al.*[8] found that adding wheels to the Calibra produced a C_D increment of 0.074.

The high value of increased drag may be attributable to interference effects; the wheel wake and flow field interfere with the body flow. Another factor, however, is that the wheels are situated in a distorted flow produced by the body. As a consequence, the air flow can meet the wheel at a yawed angle. Experimental data in Cogotti[3] show that the drag coefficient rises with yaw angle. A realistic estimate for the drag increase due to wheels for a domestic car would therefore be around 0.08.

There is some disagreement in published results for the effects of fairings on C_L, but from Scibor-Rylski[11] it appears that the lift coefficient based on total frontal area drops to about half the exposed wheel value when the top 25 per cent of the wheel is faired. From this, it might appear that for anything other than exposed-wheel racing cars the lift contribution of the wheels should be small. However, in the case of the Calibra, as reported by Mercker *et al.*,[8] it was found that the wheels generated a large amount of lift indirectly by interference with the body flow, even though the direct contribution to lift was small and negative. It is not known whether this large lift effect is common to many vehicles, but it might be expected to be so, as the Calibra is a fairly conventional car. The wide exposed tyres of racing cars produce an unwanted large contribution to (upward) lift, as described in Chapter 6.

DRAG DUE TO THE COOLING SYSTEM

The engine cooling system can produce a significant contribution to drag. The mechanisms of drag production and design principles for low drag are dealt with in some detail in Chapter 7. A few years ago, a typical value for the increase in drag coefficient due to the cooling system of domestic cars was found to be 0.04, but recent improvements in design could be expected to lower the current average to 0.03 or less.

THE RELATIVE IMPORTANCE OF THE DIFFERENT CONTRIBUTIONS TO DRAG

There is no practical means of separating and measuring the different contributions to drag, and only crude estimates for surface friction and trailing vortex drag can be made. On the basis of Carr's estimate of 0.11 for surface friction drag mentioned earlier, it may be estimated that modern cars have a surface friction drag coefficient of around 0.08.

For trailing vortex drag, Carr's k value of 0.5 and a C_L value of 0.15 may be taken as being fairly typical for a medium-size domestic car. This gives an estimated trailing vortex drag coefficient contribution of roughly 0.01. On the basis of these figures, and taking an overall drag coefficient of 0.3 as being typical for such a vehicle, the drag component breakdown may be estimated as in Table 2.3.

From the table, it can be seen that the first three items are of similar magnitude, and that although pressure drag is the largest item, it is by no means the dominant factor; streamlining the upper surfaces of the body alone will not produce exceptionally low C_D values. The trailing vortex contribution in the above case is small, although, as will be seen in Chapter 4, hatchbacks (fastbacks) can sometimes produce a significant amount of trailing vortex drag. The wings of racing cars can also be a major source of trailing vortex drag (see Chapter 6). Surface friction drag currently accounts for about a quarter of the total drag on domestic cars, and since it is dependent on the surface area rather than the frontal area, it will be proportionally higher on vehicles such as articulated trucks that are long in relation to their frontal area.

Table 2.3 Estimated contributions to the drag coefficient for a typical family car.

Contribution	C_D
Surface friction, underbody and excrescence	0.08
Normal pressure	0.10
Effects of wheels	0.08
Engine cooling system	0.03
Trailing vortex	0.01
Total	**0.30**

PRINCIPLES OF DRAG REDUCTION AND MINIMIZATION

Methods of drag reduction are dealt with in detail in the chapters devoted to specific types of vehicle. However, it is convenient at this stage to set out some general principles of drag minimization.

SURFACE FRICTION DRAG REDUCTION

Minimization of surface friction drag is partly achieved by producing a vehicle with a smoothly contoured continuous surface. Any protrusions, gaps or discontinuities in the surface will produce drag in their own right (excrescence drag), but they will also increase the surface friction drag by increasing the boundary layer thickness and turbulence.

A useful reduction in surface friction drag is possible if a large proportion of laminar boundary layer can be sustained, because a laminar boundary layer will produce less surface friction drag than a turbulent one at the same Reynolds number. For example, using simple boundary layer theory, it can be calculated that at a speed of 10 m/s, the surface friction drag on one side of a flat plate 1 m long would be 0.108 N per metre width, whereas a turbulent layer (which would have to be tripped into turbulence at the leading edge) would produce a drag of nearer 0.3 N per metre width. Maintaining a laminar boundary layer is not easy; at 100 km/h, (62 mph) transition on a smooth flat plate would occur at about 30 cm (1 ft) from the leading edge.

In order to encourage a large proportion of laminar boundary layer, it is necessary to contour the body in such a way as to maintain a favourable pressure gradient (pressure decreasing gradually along the body) as far aft as possible. Even if a large extent of laminar layer is not sustained, a favourable pressure gradient will help reduce surface friction drag by slowing the rate of boundary layer growth. Prior to the late 1980s, the rather blunt shape of the front end, accompanied by panel joint-lines and raised styling details, usually meant that there was little or no laminar layer.

The principle of contouring to produce an extended area of favourable pressure gradient started to be exploited in the design of aircraft wing sections in the late 1930s. By moving the position of maximum thickness of the section further aft, and by suitable shaping, it was found to be possible to produce a favourable pressure gradient over more than 50 per cent of the surface. This compares with around as little as 15 per cent on earlier sections. Figure 2.18 shows a comparison between a 'laminar' and an older aerofoil section. A more detailed description of low-drag aerofoils is given in Barnard and Philpott[1]. The same approach can be used in the design of racing or record-breaking vehicles, where a large proportion of the drag is due to surface friction, and the contour may be smooth and continuous enough to support a laminar boundary layer over a significant proportion of the surface. The

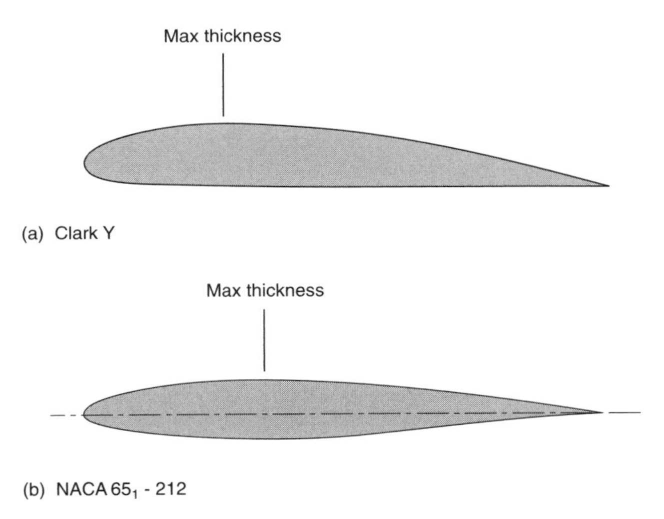

Fig. 2.18 (a) Low-drag laminar (Clark Y) and (b) an older (NACA 65$_1$-212) aerofoil section compared.

problem with applying this approach to production road vehicles, however, is that the front end would have to be radically altered to a totally smooth continuous contour with no excrescences, radiator air intake, or projecting bumper bar. In view of the relatively small advantages to be gained from such a major styling change, it is unlikely to happen on most road-going vehicles in the immediate future; there are easier methods of reducing drag.

Competition cars are often given a high polish in the hope that this will reduce the drag. In reality, polish may delay transition, but otherwise will have little effect. Laminar layers are not significantly affected by the very small level of roughness in matt-finish paint, but if the surface is too rough, the layer will simply cease to be laminar. For a turbulent layer, the drag is not affected if the height of the roughness grains is less than the thickness of the laminar sub-layer.

REDUCING THE ROUGHNESS OF THE UNDERSIDE

An area for potentially large improvement in drag is the underside. Racing car constructors pay considerable attention to the under-surface, as will be described in Chapter 6, but for normal road vehicles, designers seem almost to forget that the air is aware of the mess on the underside (Fig. 2.19) even if we cannot see it. Carr's analysis,[2] mentioned earlier, indicates that the rough underside of a typical domestic car may contribute about 0.07 to the value of the drag coefficient (20 per cent). Some progress has been made in this area since the start of the 1980s, but at the time of writing there is still plenty of scope for improvement.

Fig. 2.19 The underside of a car. Even without the exhaust and suspension elements, the underside of many cars is extremely rough.

PRESSURE DRAG REDUCTION - STREAMLINING

In order to minimize pressure drag, it is necessary to keep the flow attached as far aft as possible. Over most of the front part of the vehicle, where the cross-sectional area is increasing, the flow accelerates. From the Bernoulli relationship between pressure and speed, we can see that the increase in speed will be accompanied by a decrease in pressure. The air is flowing from a high to a low pressure, and thus there is is a favourable pressure gradient. Not only does this favour low surface friction drag, but separation is unlikely unless there is a sudden discontinuity or sharp angle in the surface. With attached flow, the front surface usually produces little or no contribution to pressure drag; indeed, because of the high flow speed and consequent low pressure, it can even provide a small negative contribution. This fact appears contrary to common sense because it means that the front of the vehicle and the windscreen are actually pulled forward, and not 'blown' backwards as would be expected; however, experimental measurements show this to be true. Once again it can be seen that intuitive ideas are often wrong because we tend to imagine air as behaving in the same way as a solid object, or at least, as a stream of solid particles.

Separation is much more likely to occur over the rear portion of the vehicle, where the air slows down and the pressure rises again (an unfavourable pressure gradient). If the pressure rise is gradual, then energy from the free stream is fed into the boundary layer fast enough to prevent separation. To inhibit separation at the rear, therefore, the cross-sectional area should decrease gradually. Even if the flow does not separate, the boundary layer thickens rapidly in an adverse pressure gradient and the pressure on the rear surfaces is reduced, thus increasing the pressure drag.

Strong adverse pressure gradients occur at sharp corners, so for low drag, a body should have a smooth outline with no sharp corners, and should taper gradually to a point at the rear. The classic example of such a 'streamlined' geometry is the teardrop shape shown in Fig. 2.20. This fulfills the requirement for a smoothly continuous surface with the cross-sectional area decreasing gradually at the rear; however, as mentioned in the discussion of surface friction drag, if the benefits of a laminar boundary layer are to be exploited, a shape giving a greater extent of favourable pressure gradient would be preferable. This requires a contour with the position of maximum cross-sectional area further aft, as with the aerofoil shown in Fig. 2.18.

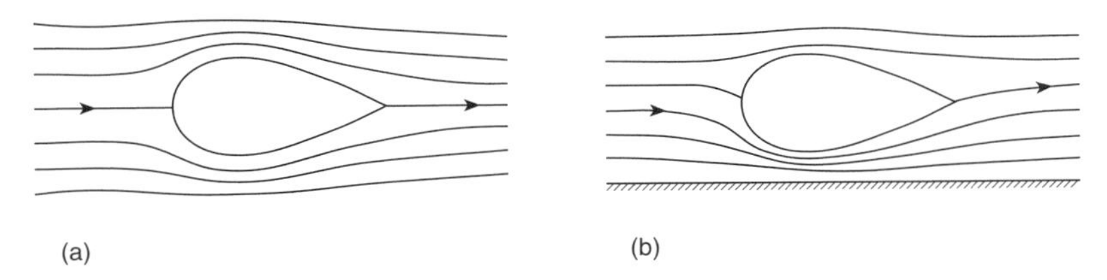

Fig. 2.20 Teardrop shape: (a) unconstrained flow; (b) close to ground.

The long tail of the teardrop shape makes it impractical for most types of road vehicle, and in Chapter 4 we describe the effect of truncating the shape with alternative rear-end compromise configurations.

FINENESS RATIO

For a teardrop shape, the minimum drag *for a given cross-sectional area* occurs when the streamwise length is about three times as great as the diameter, as shown in Fig. 2.21. Increasing the length would decrease the pressure drag, but this reduction would be outweighed by the increased surface friction drag produced by the larger surface area. Fineness ratio as shown on the figure is defined as the ratio of the body length l to its diameter d.

For a freight-carrying vehicle it would be more appropriate to minimize the drag for a given volume rather than a given frontal area. In this case the ideal shape would be thinner and longer. For a passenger vehicle, both the width and the volume are important, and the optimum shape depends on what compromises the designer chooses to make. In the post-war era, the trend has been to cut down on the cross-sectional area without reducing the useful volume. Persuading the occupants to adopt a more recumbent position and making door structures thinner are two approaches that have been used.

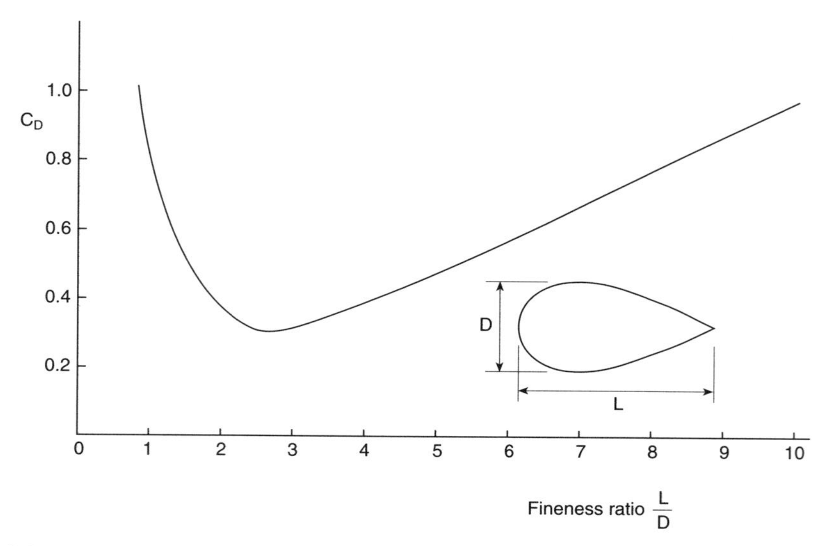

Fig. 2.21 Variation of drag coefficient with fineness ratio l/d for a body of revolution.

THE EFFECT OF GROUND PROXIMITY - HALF SHAPES

The teardrop shape represents the ideal low-pressure-drag shape for aeronautical applications, but when a vehicle is close to the ground, the situation changes. As illustrated in Fig. 2.20(b), the presence of the ground prevents the formation of a symmetrical flow pattern, and this results in an increase in drag coefficient, and the generation of a down force. Measurements by Morelli[10] of the variation of drag coefficient with ground clearance for a streamlined shape produce a relationship of the form indicated in Fig. 2.22.

If the vehicle were in direct sliding contact with the ground, then the ideal shape would be a half teardrop; the ground line then corresponds to the line of symmetry of the full body in an unconstrained air flow, as illustrated in Fig. 2.23. The half body in sliding ground contact should thus theoretically have the same drag coefficient as a full body in free air. (The drag should be half that of the equivalent full body, but the frontal area is also halved, so C_D remains the same.) This would be true in practice only if the air were not moving relative to the road surface, otherwise the boundary layer on the road would affect the flow. Note that the equivalent full body shown in Fig. 2.23 would have half the fineness ratio of the half body away from the influence of the ground; in other words, the equivalent full body is twice as fat as the actual half body. Bringing a body close to the ground therefore makes it effectively less 'fine', and this factor tends to increase the drag.

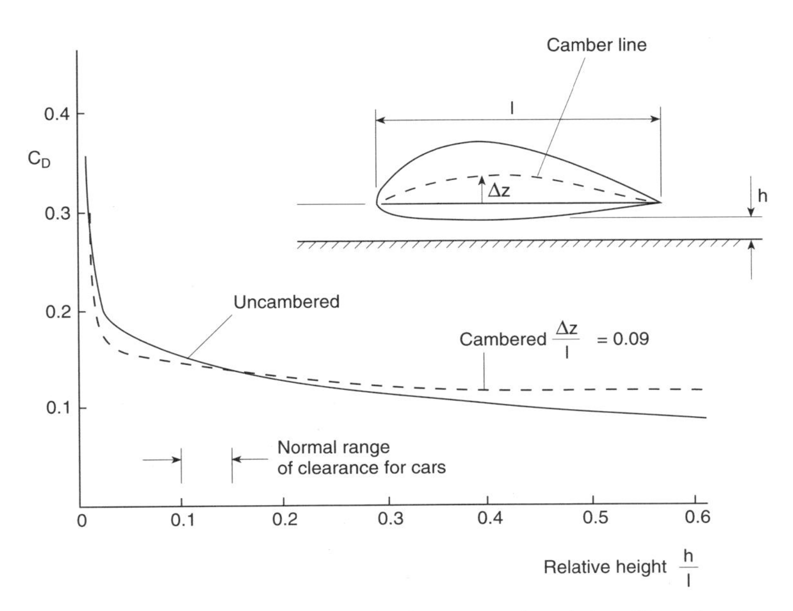

Fig. 2.22 Teardrop shape C_D against height above ground, and the effect of camber. (*After Morelli*[10])

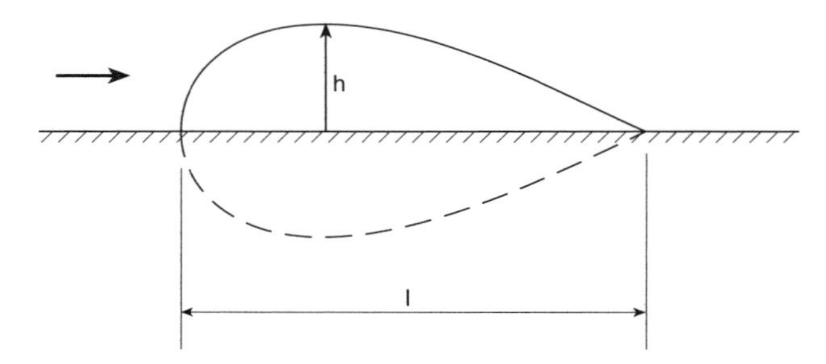

Fig. 2.23 The effect of placing a half teardrop in sliding contact with the ground. The effect is equivalent to a whole body in free flow.

Practical road vehicles are neither in sliding contact with the ground nor are they in an unconstrained air flow. The ideal shape therefore lies somewhere between the half and full teardrop, with a geometry, and particularly a degree of camber, that depends on the ground clearance. As illustrated in Fig. 2.22, the camber line is a line drawn midway between the upper and lower surfaces of the body. The **degree of camber** is the ratio of the maximum deviation of the camber line from a straight line drawn between the front and rear of the body to the body length $\Delta z/l$. Figure 2.22 shows that at typical car underbody clearance gaps, a cambered body will produce a smaller drag coefficient than a symmetrical one.

The variation of lift coefficient with ground clearance is less amenable to simple generalization than the drag coefficient. At large clearances, the venturi effect between the body and the road tends to create a down force, but at small clearances the gap becomes filled with the underbody boundary layer, thereby restricting the flow and tending to produce an upward lift. The critical height is dependent on the thickness of the boundary layer, which is in turn affected by the underbody roughness. As described previously in Fig. 2.12, the lift force also depends on how the under-surface is vented. In Chapter 6, we describe how racing cars generate a strong down force by using a small ground clearance with a rear venturi.

THE BENEFITS OF STREAMLINING

A dramatic illustration of the benefits of streamlining may be gained from Fig. 2.24, which shows a streamlined teardrop-profile fairing, compared with a circular rod that produces the same amount of drag (over a certain Reynolds number range). The drag coefficient of the circular rod can be some twenty times as great as that of the streamlined fairing. This is because the flow around the circular rod tends to separate near the position of maximum thickness.

The teardrop shape in full, half, or modified form has never proved to be the basis for a good practical vehicle shape for everyday use, and as will be shown in later chapters, satisfactory low-drag shapes can be produced without resorting to impractical 'ideal' configurations. The essence of the art of streamlining is simply to devise a geometry that does not produce strongly adverse pressure gradients or a wide wake.

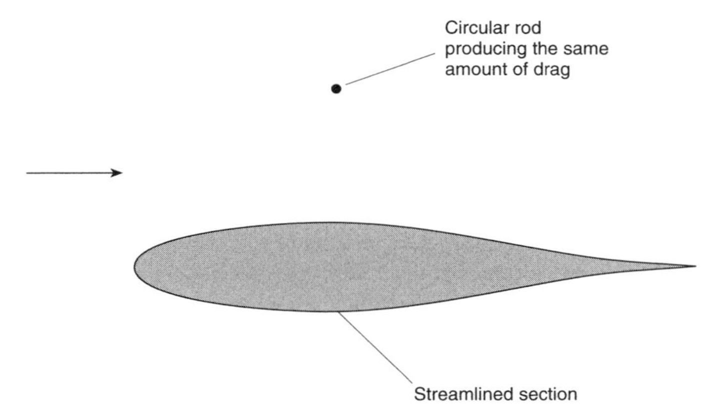

Fig. 2.24 The effects of streamlining; comparison of a streamlined fairing and a rod with the same drag.

TRAILING VORTEX DRAG REDUCTION

Aeronautical experience indicates that reducing the lift will reduce the drag, but in the case of road vehicles, the close proximity of the road strongly affects the relationship between lift and drag. The half teardrop in intimate contact with the road shown in Fig. 2.22 will produce lift because the air will accelerate over the top, with a corresponding reduction in pressure. The trailing vortices are however suppressed, because the air cannot rotate through the ground, and so there will be no trailing vortex drag. Although real vehicles are not normally in sliding contact with the road, any vortex close to the ground is distorted. A similar distortion would occur if a mirror image of a vortex was placed in proximity, as illustrated in Fig. 2.25, and in mathematical models, the presence of the ground is simulated by

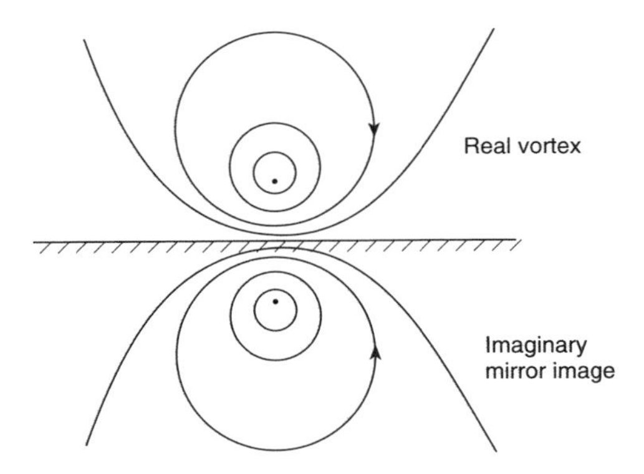

Fig. 2.25 The simulation of vortex distortion due to ground proximity, by means of an imaginary mirror image.

using a mirror-image vortex (see Houghton and Carpenter[7]). Although the mirror-image vortex does not exist in reality, placing a vortex close to the ground produces an identical effect, and it is easier to calculate and visualize how a mirror-image vortex would affect the flow than to approach the problem directly. A mirror-image vortex below the ground would rotate in the opposite sense to the real trailing vortex, and therefore, instead of drawing the air down at the rear, would tend to push it up, thereby partly cancelling the effect of the real trailing vortices; in other words, the ground tends to suppress the effects of the trailing vortex. Since road vehicle bodies are not in intimate contact with the ground, the trailing vortices will not be totally suppressed, so if there is either lift or down force, there will some associated trailing vortex drag.

Reducing the lift of a vehicle does not automatically reduce the drag; a spoiler at the end of the roof-line may kill the lift by provoking separation, but the resulting increase in wake size will increase the pressure drag, and may result in a net overall increase in drag; as a general principle, however, a low lift is essential for a low drag.

PRINCIPLES OF DRAG REDUCTION

Low surface friction drag requires a smoothly continuous surface with no sudden changes in direction, and no gaps, excrescences, or surface detail. The underside of the vehicle is just as important as the visible upper surfaces.

To minimize pressure drag it is necessary to keep the flow attached as far back as possible; this also implies continuous surface contours without sharp corners or facets. In addition, it is important that the pressure should be allowed to rise as much as possible towards the rear of the vehicle, and this means that the cross-sectional area should preferably reduce gradually towards the rear, as in a teardrop shape. The gradual reduction is necessary in order to prevent separation due to strongly adverse pressure gradients.

Trailing vortex drag can be minimized by reducing the lift (or down force).

In later chapters a description is given of how these principles are applied to the design of various different types of vehicle.

REFERENCES

1. Barnard R. H. and Philpott D. R., *Aircraft Flight*, Longman, Harlow, 1989.
2. Carr G. W., Potential for aerodynamic drag reduction in car design, in: *Impact of Aerodynamics on Vehicle Design, Proc. International Association for Vehicle Design: Technological Advances in Vehicle Design*, SP3, ed. Dorgham M. A., 1983, pp. 44–56.
3. Cogotti A., Aerodynamic characteristics of car wheels, *Impact of Aerodynamics on Vehicle Design, Proc. International Association for Vehicle Design: Technological Advances in Vehicle Design*, SP3, ed. Dorgham M. A., 1983, pp. 173–96.
4. Cresswell M. G. L. and Herz P. B., Aerodynamic drag implications of exterior truck mirrors, Paper 920204, *Proc. SAE Conf. SP-908, Vehicle Aerodynamics: Wake Flows, Computational Fluid Dynamics, and Testing*, 1992, pp. 29–33.

5. Fackrell J. E. and Harvey J. H., The flow field and pressure distribution of an isolated road wheel, in *Advances in Road Vehicle Aerodynamics*, Paper 10, BHRA, 1973.
6. Hoerner S. F., *Aerodynamic Drag*, Hoerner, PO Box 342, Brick Town N.J. 08723, USA.
7. Houghton E. and Carpenter P., *Aerodynamics for Engineering Students*, Arnold, London.
8. Mercker E., Brauer H., Berneburg H. and Emmelmann H. J., On the aerodynamic interference due to the rolling wheels of passenger cars, SAE 910311, in *Vehicle Aerodynamics, Recent Progress*, SAE SP 855, 1991.
9. Morelli A., data quoted in Cogotti A., Aerodynamic characteristics of car wheels, *Impact of Aerodynamics on Vehicle Design, Proc. International Association for Vehicle Design: Technological Advances in Vehicle Design*, SP3, ed. Dorgham M. A., 1983, pp. 173–96.
10. Morelli A., Aerodynamic basic bodies suitable for automobile applications, *Impact of Aerodynamics on Vehicle Design, Proc. International Association for Vehicle Design: Technological Advances in Vehicle Design*, SP3, ed. Dorgham M. A., 1983, pp. 70–98.
11. Scibor-Ryslki A. J., *Road vehicle aerodynamics*, 2nd edn, Pentech Press, London, 1984.
12. Stapleford W. R. and Carr G. W., Aerodynamic characteristics of exposed rotating wheels, MIRA Rep. No. 1970/2, MIRA, 1969.

CHAPTER 3

THE INFLUENCE OF AERODYNAMICS ON ECONOMY, PERFORMANCE AND ROADHOLDING

THE NEED FOR A REDUCTION IN FUEL CONSUMPTION

For many years the streamlining of anything other than racing or record-breaking cars was more a matter of styling than a real desire to reduce drag in order to save fuel. Public resistance to unfamiliar shapes made it difficult to sell vehicles which were genuinely designed for low drag. The major impetus to serious attempts at drag reduction for mass-produced vehicles came in 1973 when a group of oil exporting countries (OPEC) formed a cartel, drastically increasing the price of crude oil, and simultaneously cutting production. This move caused near panic in Western countries, which had previously taken plentiful supplies of cheap oil for granted; in Britain, fuel rationing coupons were prepared. Adding to the sense of crisis was the fact that a year earlier, an international group of experts known as 'The Club of Rome' had produced a report entitled *Limits to Growth*, which showed that the current trend of resource depletion was not sustainable, and that environmental pollution would rapidly become an overwhelming problem if left unchecked. It was readily apparent that road transport was a major factor both in terms of oil consumption and in the rising level of air pollution. Quite suddenly, therefore, designing vehicles for lower drag and hence lower fuel consumption was recognized as a matter of vital economic and environmental importance. At the time of the crisis, it was estimated (Snyder[6]) that a 1 per cent reduction in fuel consumption would save some 310 million dollars per year in the USA. Fortunately, due to public awareness of the problem, vehicle streamlining for low drag became a positive selling point, and customer resistance to the resulting unfamiliar styles was lessened.

Although the initial impetus was for improved fuel efficiency, it soon became apparent that there were other advantages to be obtained from good aerodynamic design, such as improved roadholding and performance and reduced internal noise.

FUEL CONSUMPTION, POWER, AND RESISTANCE TO MOTION

The fuel consumption of a vehicle depends on the efficiency of the engine and transmission system, and the power required to overcome the resistance to motion.

The required power is the product of the total resistance force, and the vehicle speed.

power = total resistance force × speed

At steady speed on a level road, the total resistance to motion is the sum of two separate contributions, aerodynamic drag, and tyre rolling resistance. Although this book is mainly concerned with aerodynamic effects, it is also necessary to look briefly at the contribution to total resistance from tyre rolling effects.

TYRE ROLLING RESISTANCE

The production of rolling resistance by the tyres is a complex process. Factors affecting it include

- rubber mix
- temperature
- speed
- load
- suspension rate
- suspension geometry
- tread pattern
- road surface
- inflation pressure
- driving or braking torque
- tyre geometry
- wheel width and diameter
- construction type – radial or cross-ply (bias), steel or fibre braced.

Because of the large number of factors, the accurate estimation of tyre rolling resistance is no simple matter. The SAE conference of 1977 referred to in Snyder[6] revealed a high degree of uncertainty and variability in the data available at that time. There were considerable discrepancies between laboratory and road test data, and between the results of different experimenters. The situation has improved somewhat in recent years, but the sensitivity of rolling resistance to operating conditions still provides a complication in performance calculations.

It is generally found that rolling resistance is nearly directly proportional to the download applied on the tyre:-

rolling resistance = k_r × download

where k_r is known as the rolling resistance coefficient.

The value of k_r depends on all of the factors mentioned above, but for a given tyre, the most important ones are speed, temperature, and inflation pressure. Increasing the inflation pressure reduces the rolling resistance (see Klamp[4]), but it also reduces the adhesion, so over-inflated tyres can produce a loss of braking and cornering power.

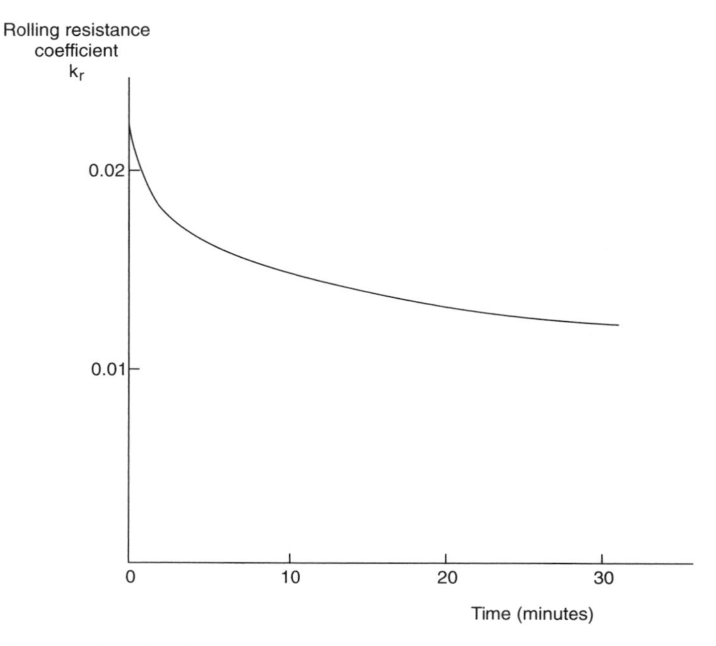

Fig. 3.1 The variation of rolling resistance coefficient with time starting from rest. (*Based on data in Hunt et al.*[2])

When a vehicle is in motion, the tyres warm up, and the rolling resistance drops, although the adhesion properties tend to improve. The temperature reached after a long period at steady speed (the equilibrium temperature) increases with speed. Figure 3.1 shows a typical curve of rolling resistance against time for steady speed driving, based on data given in Hunt *et al.*[2] Because of the time taken to warm up or cool down, the rolling resistance at any instant is dependent on the speed and conditions in earlier parts of the journey. The sensitivity of rolling resistance to temperature means that it can change with the atmospheric conditions, road surface type and dryness. It will thus vary during a journey.

ROLLING RESISTANCE AND SPEED

Figure 3.2 shows a typical curve of rolling resistance against speed, again derived from data in Hunt *et al.*[2]. The curve is based on measurements made at steady speeds, with the tyres being allowed to reach an equilibrium temperature. It does not therefore represent what would happen for a vehicle accelerating from rest. It will be seen that the line in Fig. 3.2 is nearly flat up to about 100 km/h (62 mph), but thereafter rises. Below 100 km/h the rolling resistance coefficient can be treated as a constant for most practical engineering calculations.

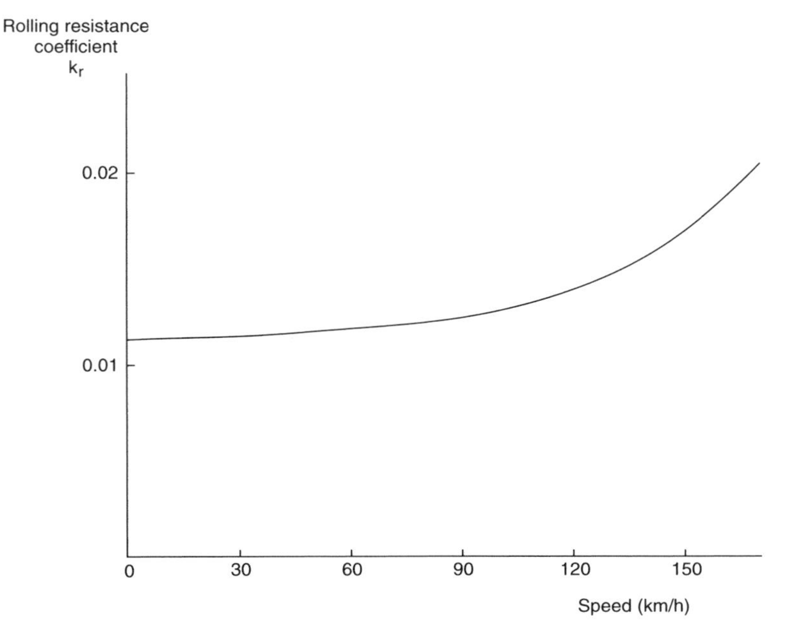

Fig. 3.2 The variation of rolling resistance coefficient with speed.
(*Based on data in Hunt et al.*[2])

Typical values for the constant k_r for family cars are

- 0.012 for radial tyres, and
- 0.01 for special 'low-loss' tyres.

The cross-ply (bias belted) tyres still used on some heavy vehicles typically have values of around 0.02. It should be noted that there are wide variations between different makes and sizes of tyre, and the values depend strongly on inflation pressure.

Radial braced tyres have almost totally displaced the cross-ply type on small and medium-sized vehicles. Thompson and Torres[9] showed fuel consumption improvements of around 5 per cent for the radial tyres relative to cross-plys in comparative driving tests. Special tyres with low rolling resistance exist, but it is difficult to combine low rolling resistance with good adhesion for cornering and braking.

The assumption of a constant rolling resistance coefficient begins to introduce significant error above 100 km/h (62 mph), so it is common practice among vehicle manufacturers to fit a linear relationship of the form

$$k_r = a + bV$$

to the curve of k_r against speed (V) in the main region of interest, which is between about 50 and 120 km/h. Since a linear relationship bears no relationship to the reality shown in Fig. 3.2, the resulting constants a and b vary significantly from one authority to another (see Phelps and Mingle[5]). Various alternative empirical relationships are used in the automotive industry.

Wheels also suffer a small aerodynamic or windage resistance torque which varies with V^2. This factor is generally small except on racing cars, and since it is related to V^2, it can be included with the aerodynamic drag.

THE RELATIVE CONTRIBUTIONS OF AERODYNAMIC DRAG AND ROLLING RESISTANCE

Figure 3.3 shows the separate effects of tyre and aerodynamic drag applied to a typical medium-sized European family car weighing 900 kg (1980 lb) with a frontal area of 1.89 m^2, a C_D of 0.35, and fitted with radial tyres.

The tyre rolling resistance coefficient k_r at any speed has been estimated from Fig. 3.2, and the aerodynamic drag is given by $\frac{1}{2}\rho V^2 AC_D$. It will be seen from Fig. 3.3 that the aerodynamic drag starts to dominate at around 60 km/h (38 mph).

AERODYNAMIC IMPROVEMENTS AND FUEL CONSUMPTION

Using the data given above, it is possible to estimate the order of fuel savings that would be produced at constant cruising speed by reducing the drag coefficient. A typical medium-sized European car with a 1.5 litre (91.5 in^3) engine and a C_D of 0.35 returns a fuel consumption figure at a constant 120 km/h (75 mph) of around 7.1 litres/100 km (39.8 mpg (miles per gallon)).

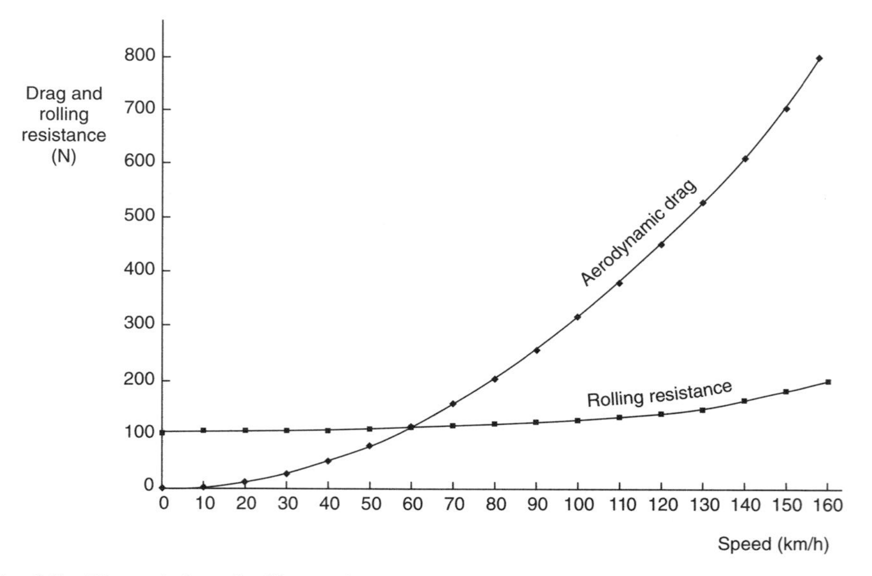

Fig. 3.3 The variation of rolling resistance and aerodynamic drag with speed for a typical medium-sized European car.

Using the same vehicle data as for Fig. 3.3, it can be calculated that a decrease in C_D to 0.25 would decrease the aerodynamic drag by 128 N, and would thus reduce the total resistance (rolling + aerodynamic) from 560 N to 432 N at 120 km/h: a 23 per cent decrease. If all other factors such as engine efficiency were unaltered, this would represent a similar reduction in fuel consumption, which would thus be 5.47 litre/100 km (51.6 mpg), a worthwhile improvement.

Unfortunately the assumption of unaltered engine efficiency is not usually valid. Engines generally have a higher efficiency when working hard with a high mean effective pressure. In the low-drag vehicle, the engine will be working less hard, and the efficiency will consequently be lower. To realize the full potential for fuel saving it would be necessary to raise the gear ratio, or to use a smaller engine. Both options would unfortunately have an adverse effect on the performance in terms of acceleration. The evidence is that the public is still demanding better rather than inferior performance.

THE EFFECT OF THE ATMOSPHERIC WIND ON THE DRAG COEFFICIENT - WIND AVERAGING

In the simple estimate of fuel consumption made above, the value of C_D used was that for no atmospheric wind, and no account was taken of the effect of any tailwind, headwind or side-wind effects. In terms of long-term fuel consumption assessment, however, it is necessary to appreciate that on most journeys there will be a certain degree of atmospheric wind. Buckley *et al.*[1] have estimated that the mean wind speed during the normal driving hours in the USA is 2.05 m/s (6.8 ft/s) at 2.1 m (7 ft) height above the road.

It might be thought that on average the wind would have a swings and roundabouts effect, in that it would impede the vehicle when blowing from the front quarter, and help it by an equal amount when blowing from behind. There are, however, two reasons why this is not the case. The first is that the drag is related to the square of the relative air speed, so the increase in drag due to a headwind is proportionally greater than the decrease due to a tailwind. Taking for example, the case of a vehicle with a C_D of 0.35 and an area of 1.89 m^2 travelling at 20 m/s in a wind blowing at 10 m/s, the drag figures in Table 3.1 are obtained. It can be seen that the average drag for a combination of headwind and tailwind conditions is higher than for the still air case.

Table 3.1 Headwind and tailwind effects for a car travelling at 20 m/s. The average drag value for a combination of tailwind and headwind cases is greater than the drag in the still air condition.

Wind conditions	Relative speed V (m/s)	V^2	Drag (N)
No wind	20	400	162
Headwind 10 m/s	30	900	365
Tailwind 10 m/s	10	100	41
Average drag (tailwind + headwind)			203

The second factor influencing the drag in windy conditions is that the C_D tends to rise when the resultant relative air flow is yawed relative to the direction of vehicle motion, as discussed in the previous chapter. When there is a cross-wind component, therefore, not only will the relative speed increase, but the drag coefficient will rise due to the effect of yaw. Buckley *et al.* found that for a large truck in an 11 km/h (7 mph) wind at an 88 km/h (55 mph) driving speed, the drag increase was positive for yaw angles up to around 120°; that is, even with the wind blowing from the rear quarter.

To take account of such wind effects, it is possible to determine a so-called 'wind-averaged' drag coefficient based on meteorological data, and the drag/yaw characteristic of the vehicle. Buckley *et al.*[1] give a suitable expression involving wind speed and direction probability functions. Using this expression and the data for large trucks in the USA, it was found that at 100 km/h (62 mph), the wind-averaged C_D was some 10% higher than the still air or wind-tunnel value.

STANDARD DRIVING CYCLES

The calculation of fuel consumption improvement given earlier was for the simple case of cruising at constant speed. Most road vehicle journeys involve a great deal of speed change and often some stopping and starting. Under these conditions most of the energy used to accelerate the vehicle is dissipated as heat during braking, so fuel consumption cannot be calculated simply on the basis of the amount of rolling and aerodynamic resistance. In order to obtain a rational basis of comparison of the fuel consumption of vehicles on typical journeys, standard tests have been devised. Several countries have their own national tests, but the most commonly quoted standards are the European ECE and American EPA. Both standards have tests for simulated urban and highway driving conditions and a standard mixture of highway and urban. Most tests are based on schedules of speed changes that are intended to be typical of normal patterns of use, but the European highway tests are represented by simple constant speed driving.

The schedules are modified from time to time, but those current at the time of writing are illustrated in Fig. 3.4. The European urban cycle is based on dynamometer tests rather than real driving, and does not directly take account of the aerodynamics of the vehicle. Even the actual weight is not used, merely the central value in a series of weight bands. Changing the weight of the vehicle will only affect the tests if the change is sufficient to move from one band to another. The American EPA urban cycle can be based on dynamometer tests or road testing. The EPA schedules for highway driving are also shown in Fig. 3.4. The European highway driving schedules are simply a constant 90 km/h (56 mph) and a constant 120 km/h (75 mph).

- The EPA mixed or composite schedule is 55 per cent urban + 45 per cent highway.
- The ECE 'Euromix' is $\frac{1}{3}$ urban + $\frac{1}{3}$ 90 km/h + $\frac{1}{3}$ 120 km/h

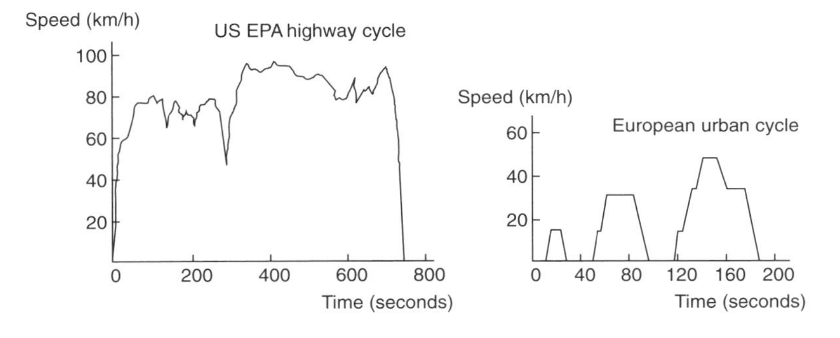

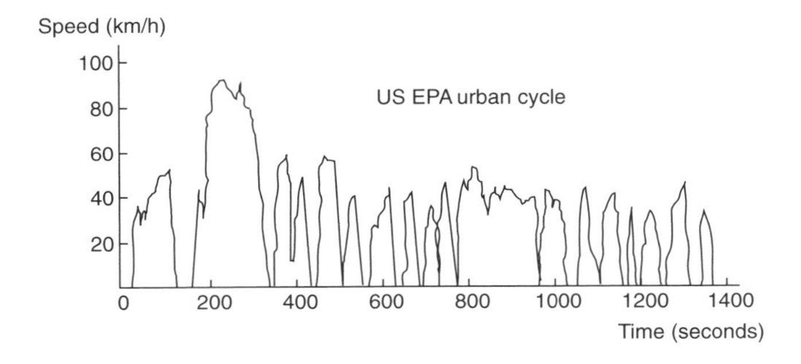

Fig. 3.4 The EPA and European driving schedules.

Changes aimed at making the European tests more representative are to be introduced; these include a cold start for the urban schedule.

Sovran and Bohn[7] give details of calculations for the EPA cycle. Two interesting facts emerge from their analysis, namely that for a typical car driving the EPA urban cycle, 40.7 per cent of the time is spent with the throttle closed, and 34 per cent of the energy input from the engine is dissipated in braking.

A detailed experimental investigation of the effect of aerodynamic drag reduction on fuel consumption was conducted at MIRA by Stapleford[8] in the 1970s using a small saloon car with a basic (zero yaw) drag coefficient of 0.45, which was fairly typical for such a vehicle at the time. With progressive modifications to the bodywork and cooling system, the drag coefficient was reduced to 0.316. Table 3.2 shows Stapleford's computed figures for fuel consumption on the EPA cycles. The improvement in fuel consumption is quite marked for the highway and composite cycles, and since a drag coefficient of around 0.32 is fairly typical of recent designs, the data give a good indication of the likely impact that aerodynamic design has had on domestic vehicle fuel consumption since the late 1970s.

Nowadays the cruising speed of all sizes of commercial vehicle is restricted primarily by speed limits rather than lack of power, and the influence of

Table 3.2 Comparison of computed fuel consumption for a small saloon car driving the EPA cycles, for baseline and low-drag modified vehicles.

EPA cycle	Fuel consumption (litres/100 km)		Percentage improvement
	Baseline (C_D at 0° = 0.45)	Modified vehicle (C_D at 0° = 0.316)	
Urban	6.88	6.42	7
Highway	6.37	5.25	18
Composite	6.64	5.90	11

(Based on data in Stapleford[8])

aerodynamic drag on their fuel consumption is just as significant as for domestic cars. Details of fuel consumption reductions for commercial vehicles due to aerodynamic improvements are described in Chapter 5.

AERODYNAMIC INFLUENCES IN HYBRID VEHICLES

Due to the influence of legislative pressure to introduce electric or low-emission vehicles in urban areas, there is currently a growing interest in hybrid propulsion systems where the installed engine power is just sufficient to maintain cruise speed up a moderate incline, and the short bursts of energy required for acceleration are provided by a storage system such as a battery or flywheel (Jefferson and Barnard[3]). The use of energy storage also allows the energy which is normally dissipated as heat during braking to be recovered. Because of the improved overall efficiency in a hybrid system, the aerodynamic drag assumes even greater importance in achieving low fuel consumption.

INFLUENCE OF AERODYNAMIC DRAG ON ACCELERATION

For an accelerating vehicle, Newton's laws give

$$ma = F_t - F_r - \tfrac{1}{2}\rho V^2 A C_D - mg \sin\theta \qquad [3.1]$$

where m is vehicle mass
a is the acceleration
F_t is the tractive force available *at the wheels*
F_r is the rolling resistance
θ is the angle of inclination of the road

On a level road, the $mg \sin\theta$ term disappears.

The tractive force F_t at the wheels depends either on the power available at the tyre–road interface P_a, and the speed V, or on the maximum amount of thrust that the tyre can transmit without slipping. If the tractive force is limited by the power available at the driving wheel–road interface

$$F_t = P_a / V \qquad [3.2]$$

Note that the power available at the driving wheels is much less than the steady-speed power output from the engine, because a significant amount of energy goes into accelerating the rotating masses of the engine and transmission. In a racing car, this can be as much as 40 per cent of the power produced within the engine. Friction losses in the transmission system also reduce the amount of power available at the road.

Figure 3.5 shows the influence of drag coefficient on acceleration, based on data for a typical European car weighing 1000 kg. In the popular press, acceleration times are quoted as the time for 0–100 km/h (or 0-60 mph), and it will be seen that a change in C_D from 0.45 to 0.35 provides an improvement of 0.5 s in the acceleration time to 100 km/h. This may not seem much, but would be a noticeable improvement. Reducing C_D to a very low value of 0.25 would make a further 0.5 s improvement.

In Europe, it is the time from 0–120 km/h (0–75 mph) that is probably more useful, as 120 km/h represents a typical motorway cruising speed in many European countries. For the acceleration to 120 km/h, the car with a C_D of 0.25 is three seconds faster than the vehicle with a C_D of 0.45. For 0–140 km/h the effect of C_D becomes even more important, partly because the aerodynamic drag increases with the square of the speed, and partly because the high-drag vehicle is nearing its maximum attainable speed.

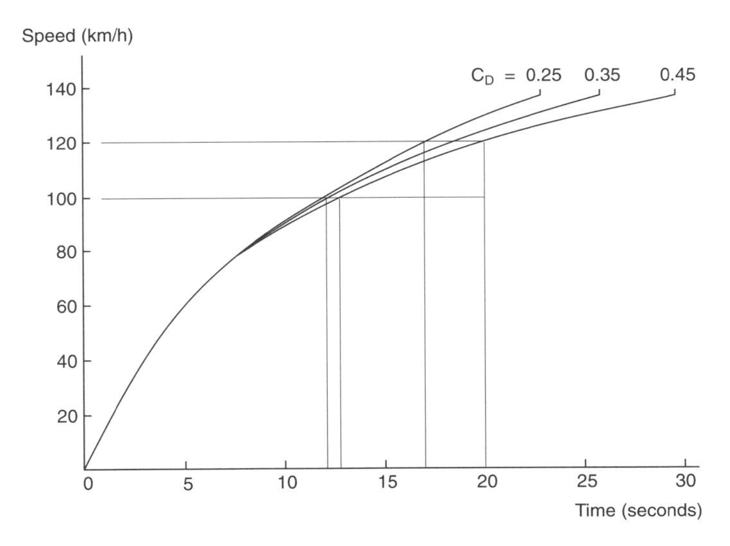

Fig. 3.5 Effect of drag coefficient on acceleration.

INFLUENCE OF AERODYNAMIC LIFT OR DOWN FORCE ON ACCELERATION

At low speeds, the tractive force available at the road is limited not by the power available, but by the adhesion: the amount of tangential force that the wheels can apply to the road without spinning. This is related to the down load on the wheel by

$$F_t = k_s\,(\text{max}) \times \text{down load}$$

Where k_s (max) is the maximum slip coefficient, which is effectively the maximum coefficient of friction between the road and the tyre. The reason that the coefficient is called the slip coefficient is that under hard acceleration, the peripheral speed of the tyre is greater than the road speed, the difference being known as the degree of slip. It should be noted, however, that in the vicinity of the small area of contact between the road and the tyre, the contact patch, the tyre is not actually sliding, but merely distorting to take up the difference in speed. Once the rubber does start sliding relative to the road, the adhesion force starts to decrease. A similar situation applies on heavy braking.

Increasing the load by making the vehicle heavier will not improve the acceleration, because although the tractive force can be increased, the inertia of the vehicle will increase in equal proportion due to the extra mass, so correspondingly more tractive force will be required. If the down load is increased by purely aerodynamic means, however, there is no increase in inertia, and the acceleration will improve if sufficient power is available. Once aerodynamic down load is introduced, the acceleration can be increased by reducing the vehicle mass.

Most road-going vehicles are only adhesion-limited at very low speeds, where the effect of aerodynamic down load is negligible, but high-powered track-racing cars can be adhesion-limited at high speeds, and by preventing wheel spin or skidding, aerodynamic down load can produce a significant improvement in both acceleration and braking force.

AERODYNAMIC DRAG AND MAXIMUM SPEED

The tractive power required to overcome the resistance is given simply by

$$\text{power} = \text{resistance} \times \text{speed}$$

Figure 3.6 shows a plot of power against speed using the same data as for Fig. 3.5 but for three different values of C_D.

If the vehicle in question has a maximum power available at the wheels of 52 kW (70 bhp), the maximum speed rises from 160 to 173 km/h (100 to 108 mph) for a drag coefficient reduction from 0.45 to 0.35. A further C_D reduction to 0.25 raises the maximum speed to 191 km/h (120 mph). It should be noted, however,

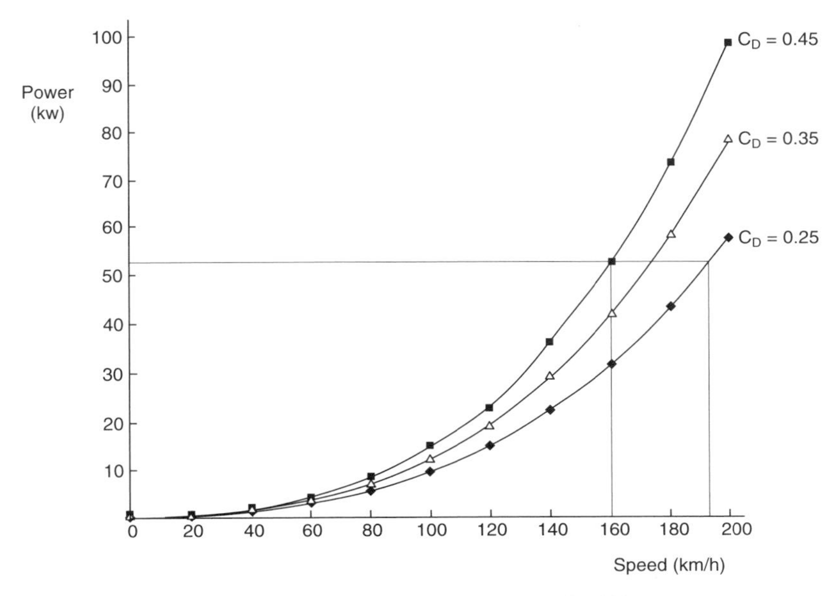

Fig. 3.6 Total power required against speed for a medium-sized European car.

that these improvements will be possible only if the final drive gear ratio is changed so that the maximum power is developed at the maximum speed. Domestic vehicles are not in fact optimized for maximum speed, since that would be well above the legal speed limit, and the acceleration is improved by gearing for maximum power at a lower speed.

CORNERING FORCES

To explain road vehicle cornering characteristics, it is necessary to return briefly to the subject of tyres. When a vehicle is cornering, the direction in which the wheels are pointing is different from the direction of motion, as illustrated in Fig. 3.7. The angle between the wheel orientation and the direction of motion is called the **slip angle**. At first sight, it might be thought that this implies that the tyre is actually sliding slightly relative to the road. In fact, due to the elastic properties of the tyre this is not so; the sideways motion is accommodated by the tyre distorting elastically in the region of the contact patch. In this condition, the tyre can contribute to the side or cornering force necessary to provide the centripetal acceleration. The tyre side force increases almost linearly with slip angle up to a certain point, beyond which the rate of increase of force begins to fall off and a maximum side force is eventually reached.

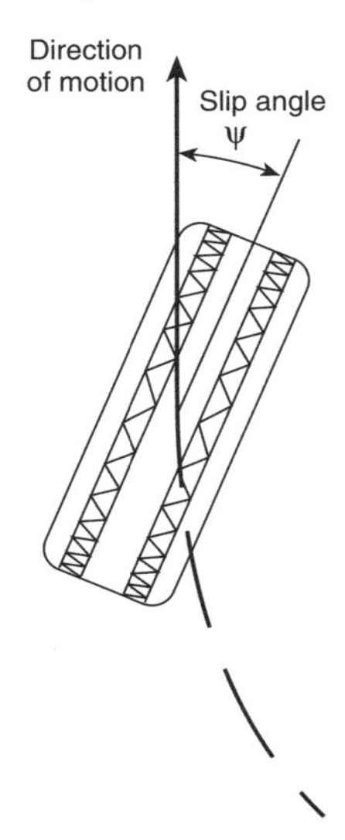

Fig. 3.7 Slip angle
Under cornering, the tyre does not point in the exact direction of motion. The angle is called the slip angle, but note that this does not mean that the tyre has to slide.

The tyre side force depends on both the slip angle and the down load on the tyre. The maximum side or cornering force S_c that the tyre can produce is given by the expression

$$S_c = k_c\,(\text{max}) \times \text{down load}$$

where k_c (max) is the maximum lateral adhesion coefficient, or maximum cornering coefficient. The value is normally similar to the longitudinal maximum slip coefficient k_s (max), and is dependent on many factors including contact patch shape, rubber compound mix, temperature, road surface, inflation pressure, and tread. The coefficient k_c (max) decreases with load, but as a first approximation for simple estimates it may be treated as a constant.

The maximum side force obtainable is reduced if the tyre is also generating either a longitudinal thrust or a braking force. Sliding will take place when the *resultant* of the longitudinal thrust and the side force exceeds the maximum tangential force that the tyre can provide.

The maximum side force that can be generated on an actual vehicle is not as high as that for an isolated wheel, because the suspension geometry and the effects of roll and mass distribution all have a strong influence; it would be unusual for both the front and rear tyres to reach their maximum cornering ability simultaneously. The maximum cornering force will, however, still be roughly proportional to the overall down load, though modified by the load distribution and the suspension characteristics. On a racing car with carefully optimized suspension set-up and good driver control, the constant of proportionality can approach the k_c (max) value.

THE EFFECT OF AERODYNAMIC DOWN FORCE ON CORNERING

Since the tyre maximum side force or lateral adhesion is roughly proportional to the down load on the tyre, adding aerodynamic down force to the weight component will obviously improve the adhesion. The force available for centripetal acceleration is therefore greater, and if the down force is distributed fairy evenly between the four wheels, the vehicle will be able to negotiate a given radius of turn at a higher speed. Conversely, if the vehicle produces positive (upward) lift, the cornering ability is adversely affected.

Simply increasing the weight of the vehicle to improve the grip will not improve the cornering ability – a fact which many people find hard to believe. The maximum grip or side force produced by the wheels increases roughly linearly with weight, but centripetal force required $(W/g) \times (V^2/r)$ also increases linearly with the mass (weight/g), so the effect of weight cancels out. Note that in reality, there may be some second-order effects, and a weight increase can actually reduce the cornering ability.

The addition of aerodynamic down force does produce an improvement in cornering, because the increased down load on the tyres increases the cornering force available without increasing the mass. As the mass does not change, a correspondingly higher centripetal acceleration (V^2/r) may be employed. This means that a higher cornering speed may be used.

The importance of aerodynamic down force in racing cars is dealt with further in Chapter 6. For road-going vehicles, aerodynamic down force is not normally employed, but reducing the natural tendency to produce positive (upward) lift is advantageous from the point of view of both roadholding and induced drag.

THE BALANCE OF LIFT FORCES BETWEEN FRONT AND REAR WHEELS

The position of the centre of lift is important, because together with the centre of gravity position, it determines the distribution of down force between the front and rear wheels, and thus affects the understeer/oversteer characteristics of the vehicle, which in turn determine the ultimate cornering ability. Applying large amounts of down force to the rear wheels will not improve the cornering if the front wheels have insufficient adhesion to steer the vehicle into the corner. On racing cars, where very large amounts of aerodynamic down force are generated, the handling and cornering ability are critically dependent on the front:rear lift distribution. Further information on vehicle cornering and handling may be found in Wong.[10]

IMPLICATIONS FOR AERODYNAMIC DESIGN

From the foregoing, it can be seen that reducing the aerodynamic drag can produce a major improvement in fuel consumption even in urban driving conditions. Drag reduction can also produce worthwhile gains in the performance in terms of both top speed and acceleration.

Aerodynamic lift, which was at one time hardly considered, affects the cornering, braking and acceleration. Lift also influences the stability and and general handling characteristics, as described in Chapter 10. Traditional vehicle shapes tended to produce quite large lift coefficients, but in the next chapter it will be shown that the lift on domestic cars can be reduced by quite straightforward design principles.

REFERENCES

1. Buckley F. T., Marks C. H. and Walston W. N., A study of aerodynamic methods for improving fuel economy, US National Science Foundation, final report SIA 74 14843, University of Maryland, Dept. of Mech. Engineering, 1978.
2. Hunt J. D., Walter J. D. and Hall G. L., The effect of tread polymer variations on radial tire rolling resistance, *Proc. SAE Conf. P-74, Tire Rolling Losses and Fuel Economy*, 1977, pp. 161–8.
3. Jefferson C. M. and Barnard R. H., Emission reduction by kinetic energy recuperation, *Proc. 24th ISATA Conference, Electric and Hybrid Vehicles*, Florence, May 1991.
4. Klamp W. K., Power consumption of tires related to how they are used, *Proc. SAE Conf. P-74, Tire Rolling Losses and Fuel Economy*, 1977, pp. 5–11.
5. Phelps R. E. and Mingle J. G., Pavement and tire rolling resistance for vehicle energy prediction, *Proc. SAE Conf. P-74, Tire Rolling Losses and Fuel Economy*, 1977, pp. 123–32.
6. Snyder R. H., Keynote address, *Proc. SAE Conf. P-74, Tire Rolling Losses and Fuel Economy*, 1977.
7. Sovran G. and Bohn M., Formulae for the tractive-energy requirements of vehicles driving the EPA schedules, SAE 810184, 1981.
8. Stapleford W. R., Aerodynamic improvements to the body and cooling system of a typical small saloon car, *Proc. 4th Colloquium on Industrial Aerodynamics*, Fachhochschule Aachen, June 18–20, 1980, pp. 121–32.
9. Thompson G. D. and Torres M., Variations in tyre rolling resistance, a real world information need, *Proc. SAE Conf. P-74, Tire Rolling Losses and Fuel Economy*, 1977, pp. 49–63.
10. Wong J. Y., *Theory of Ground Vehicles*, Wiley, New York, 1978.

CHAPTER 4

THE AERODYNAMIC DESIGN OF FAMILY CARS

For a family car, the aerodynamic considerations are, in rough order of priority, drag, cross-wind stability, cooling, ventilation, and aerodynamic noise. Some caution is necessary when assigning priorities, however, as any of the above factors can be dominant in a particular vehicle. For example, the Ford Sierra (Fig. 4.1) in its earliest form suffered from criticism for alleged instability in a cross-wind. Just how significant this instability was is disputed, but nevertheless the adverse publicity caused manufacturers to realize that stability could be a critical factor in the sales battle. In this chapter the main focus is on aerodynamic drag; the other aspects are considered in Chapters 7, 8 and 10.

Fig. 4.1 The Ford Sierra
One of first of the newer generation of aerodynamically styled cars, with a C_D of around 0.34.

THE EVOLUTION OF THE LOW-DRAG CAR

This book is not intended to provide a history of aerodynamic styling; for that, the reader is directed to Ludvigsen,[19] McDonald,[21] Kieselbach,[14,15] Koenig-Fachsenfeld,[17] Howard[9] and Klemperer,[16] but some historical background is useful because many potentially good ideas have been abandoned simply because, at the time, they were not in accord with the public perception of how a car should look. An interesting example of this is the Rumpler limousine built in 1921 (Fig. 4.2). This vehicle was unusual in that it was based on an aerofoil shape in the plan view, whereas most early low-drag designs were based on a teardrop or aerofoil shape for the side profile. When tested in the VW wind-tunnel in 1983 it was found to have the surprisingly low C_D value of 0.28, better than almost all current saloon cars. A modern apparent pastiche of this design was shown as a concept car at the 1994 Birmingham Motor Show (Fig. 4.3).

In classical times, it was believed that there were fundamental rules of aesthetics, similar to the laws of mathematics. A more recent concept, however, is that all perceptions of art and aesthetics are relative, and depend on our experience; in crude terms, people like what they are used to. This certainly seems to apply to the public's attitude to car styling. Attempts at selling aerodynamically derived forms have only been acceptable when the overall shape does not diverge significantly from the fashion of the time. Many perfectly practical low-drag designs were produced in the 1920s and 1930s, but it was not until the late 1970s that aerodynamic considerations started to dominate the styling.

Fig. 4.2 The Rumpler limousine of 1921, which was found to have a C_D of only 0.28 when tested in the VW wind-tunnel in 1983.
(*Photograph courtesy of Volkswagen AG, and Deutsche Forschungsanstalt für Luft- und Raumfahrt*)

Fig. 4.3 The 1994 'Tweaki' concept car, which bears a resemblance to the Rumpler limousine, showing how old ideas can resurface. Judging by the large side area, the name might well describe its cross-wind handling characteristics.

ALTERNATIVE APPROACHES TO LOW-DRAG DESIGN

There are two alternative approaches to low-drag design, popularly referred to as 'streamlining'. The radical method is to start with a simple low-drag shape such as the near half-body shown in Fig. 4.4, and to try to evolve a practical vehicle from it. Figure 4.5 shows one early attempt described later. The alternative conservative

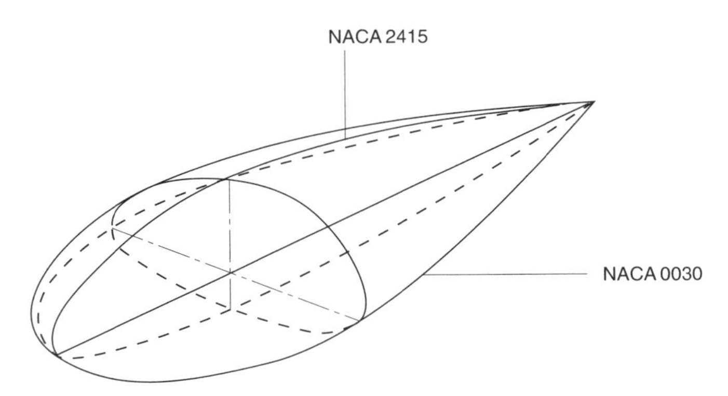

Fig. 4.4 Scibor-Rylski's ideal semi-half-body shape based on two standard aerofoil profiles.

Fig. 4.5 The Schlör/Göttingen Institute low-drag research vehicle.
(*Photograph courtesy of Volkswagen AG, and Deutsche Forschungsanstalt für Luft- und Raumfahrt*)

Fig. 4.6 The Citroën XM
Though bearing a family resemblance to earlier aerodynamically styled Citroën cars, it is not substantially different in appearance from vehicles developed from more conservative conventional forms.

approach is to start with a conventional vehicle shape, and to try to reduce its drag by gradual modification. In commercial terms, it is the latter technique that has generally been the more successful, though even here, little real progress was made amongst the major manufacturers until the 1980s. Ultimately the two methods tend to converge towards similar solutions. The Citroën XM shown in Fig 4.6 is an example of a vehicle that illustrates this point. There is a distinct family

Fig. 4.7 Citroën has produced a series of stylish aerodynamically optimized shapes since the war, starting with the DS in the 1950s, which follows the design principles of Jaray.

resemblance to earlier Citroën vehicles such as the DS (shown in Fig. 4.7), which was derived from ideal aerodynamic shapes, and yet the XM appears quite similar to mainstream contemporary European cars which have evolved gradually from the box-like shapes of horse-drawn carriages.

THE RADICAL AERODYNAMIC APPROACH

One of the best-known early champions of the radical approach was Paul Jaray, an Austrian engineer who initially worked for Count von Zeppelin on airship design. After the 1914–18 war, aeronautical development was restricted in Germany, so he turned his attention to the design of low-drag motor vehicles. Using the large wind-tunnel that had been built for airship development, Jaray evolved a number of vehicle shapes based on combinations of two- and three-dimensional streamlined elements (see Koenig-Fachsenfeld[17]). Some of Jaray's designs were tested in wind-tunnels by Klemperer[16], and C_D values as low as 0.29 were obtained, comparable with figures for the best of today's cars.

Jaray's ideas were taken, and developed into actual vehicles by manufacturers in both Europe and the USA; the list of companies includes such well-known names as Audi, Chrysler and Tatra. Unfortunately, despite the widespread commercial interest, few of the Jaray-inspired designs were financially successful. One exception was the Czechoslovakian Tatra V8 of 1934 (Fig. 4.8). Though not a direct Jaray design it was nevertheless based on his ideas. This vehicle had a C_D value of 0.34, which was remarkably low for its time. Developments of the design remained in production for many years, but this may be partly attributed to the fact that the Tatra company was run as a communist state monopoly in the post-war years, and thus public attitudes

Fig. 4.8 The Jaray-inspired Tatra V8 of 1934, based on a combination of streamlined forms. The styling was advanced for its time and was retained for many years.

Fig. 4.9 The Chrysler Airflow
A commercially unsuccessful pre-war American venture into streamlining. Though not actually designed by Jaray, it was sufficiently close to his patents to require payment of licence fees.

to styling were of no consequence. The Chrysler Airflow of 1934 (Fig. 4.9) was also not an actual Jaray design, but was sufficiently close to his patents to require payment of licence fees. This car is much more popular now as a collector's item than it was in its time. The commercial failure led to an aversion to genuine aerodynamic styling

Fig. 4.10 The SAAB 99 saloon was popular in Scandinavia. Initially fitted with a three-cylinder two stroke, it was later upgraded with a V4.

amongst the major American manufacturers that was not overcome until the 1980s, when low-drag fuel-efficient foreign imports became a major threat.

Apart from Tatra, there were other occasional commercial successes of designs following the Jaray approach, particularly in the post-war period. The French Citroën DS shown in Fig. 4.7 was popular in its country of origin; being both elegant and 'different', it was perceived as typifying national style in a country that was anxious to retain its cultural identity. The Swedish SAAB (Fig. 4.10) became the Scandinavian equivalent of the Volkswagen Beetle.

MORE EXTREME RADICAL SOLUTIONS

The Jaray approach still makes some concessions to conventionality in that the aerodynamic shapes are built up to produce vehicles similar in overall layout to more traditional cars. There are usually recognizably separate enclosures for accommodating occupants, engine and wheels, at least in vestigial form. A more radical solution is to start with a simple idealized modified teardrop or aerofoil shape. As was explained in Chapter 2, due to the proximity of the ground, the 'ideal' form for a road vehicle is neither an axisymmetric teardrop nor a half teardrop, but a cambered version, slightly flattened on the underside, with the optimum geometry being dependent on the ground clearance. A further complication arises from the fact that the cross-section will need to contain at least two people seated side by side, so a circular or semicircular section is unsuitable. Also, if excessive frontal area is to be avoided, the surrounding shell needs to fit the occupants quite closely. With these conditions imposed, there is no longer a single 'optimum', and the best shape for low drag will depend on the ground clearance and the degree of tailoring required to accommodate the occupants and internal components.

In the past, designers of ultra-low-drag vehicles have taken as their point of departure a simple teardrop-like form based on intuitively selected combinations of aerofoil profiles or simple curves generated by mathematical functions. The resulting shapes were then refined by means of wind-tunnel tests. An early example of the use of this approach is shown in Fig. 4.5. This experimental vehicle was developed at the Göttingen Institute in Germany in the late 1930s (see Hansen and Schlör[8]). The centre-line profile was based on two Göttingen aerofoil sections. The full-size vehicle had a drag coefficient of just under 0.19. Fuller descriptions of this car may be found in Koenig-Fachsenfeldt[17] and Hansen and Schlör[8]. The advantages of the low drag coefficient were to some extent offset by a fairly large frontal area, which was necessitated by the use of fully enclosed wheels both front and rear. The design also featured an impractical or at least unattractive interior arrangement which involved a central driving position.

Similar low-drag basic shapes have been developed by other researchers including Kamm and Goudert[13] and Scibor-Rylski[24]. Scibor-Rylski's geometry is shown in Fig. 4.4. In realistic proximity to a fixed groundboard (as opposed to a moving road), this body (with no wheels) gave a C_D value of 0.07. Simple shapes such as this are not generally used as a point of departure for the design of mass-produced family cars, but they have been employed in the development of record-breaking vehicles (Fig. 4.11).

Fig. 4.11 Streamlined modified half-body
Donald Campbell's Bluebird, which took the land speed record to 648 km/h (403 mph) in 1964. (*Photographed at the National Motor Museum, Beaulieu*)

The problem with such highly ideal designs is that although it is not difficult to produce a basic shape with a C_D in the range 0.05 to 0.07, adding wheels and other essential elements generally results in C_D values in excess of 0.15 even for record-breaking vehicles. One of the lowest drag designs, even amongst the very special category of record breakers, was the Ford UFO2 ultra economy vehicle which won the Shell Mileage Marathon in 1984. This little car had a C_D of only 0.113, which helped it to achieve 3803 mpg using a 15 cc engine.

In recent years, many low-drag concept cars representing almost practical vehicles have been produced, including the Ford Probe series (not to be confused with the later production Probe model). The 1985 Probe V concept car had a drag coefficient of only 0.137. The near practicality of such vehicles is deceptive, however, and Buchheim *et al.*[2] show that by the time the design has been modified to suit manufacturing, legal and safety requirements such as the inclusion of wing mirrors and the ability to fit snow chains on the wheels, the drag coefficient can easily rise to 0.3. Nevertheless, the General Motors Calibra (Fig. 4.12) with its C_D of 0.26 shows what can be achieved in a production vehicle, albeit within the relaxed constraints of a sports car.

The main requirement of a basic low-drag shape is that it should maintain attached flow over most of the surface, and there is little penalty in moving from simple teardrop or aerofoil-based body shapes to forms that are more readily adaptable to practical constraints. Morelli[22] describes how a relatively conventional-looking car, commonly known as the Pininfarina 'Banana' (Fig. 4.13) was derived from a more complex cambered low-drag shape. The basic wheelless body had a C_D of 0.073, similar to that of Scibor-Rylski's simpler shape. Once the wheels had been added, and the rear truncated to produce a practical configuration, the drag coefficient had increased to nearly 0.16 despite the careful integration of the wheels

Fig. 4.12 The GM Calibra, which has a genuine C_D of 0.26. (*Photograph courtesy of Jeff Harkman*)

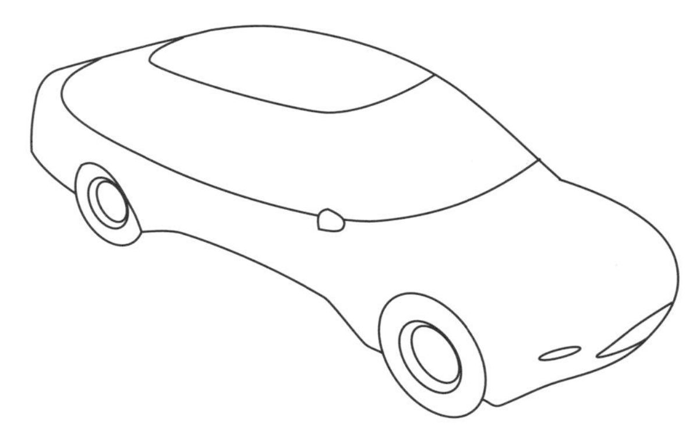

Fig. 4.13 The Pininfarina 'Banana'
A styling concept car designed for very low drag.

(as is evident in Fig. 4.13). By the time the car had been developed to meet practical and legal requirements, the drag coefficient had risen to above 0.2, and a fully engineered vehicle of this type would undoubtedly end up with a higher figure.

RELAXING THE CONSTRAINTS, TRUNCATING THE REAR

One of the main practical objections to the teardrop-based form is the extended tail, which is not a useful shape or location for any major component or passenger, and produces a long vehicle. An obvious compromise is to cut the tail down, but the form which this truncation takes is important. From a styling point of view, the temptation is to curve the rear down smoothly, as in a modern hatchback. Unfortunately, the resulting increase in camber results in a rise in lift with a consequential increment in trailing vortex drag. Figure 4.14 illustrates some of the results of a study on rear-end truncation obtained by Lay[18] of the University of Michigan in the early 1930s. A truncation in the form of a curved back was a popular style in American cars of the period (Fig. 4.15), but as may be seen from Lay's data, the drag coefficient produced is much larger than for the basic long-tail form. At about the same time, however, it was found that if a good initial taper in both profile and plan views was used, and the rear end was cut off square so that no increase in camber resulted, then the drag penalty due to truncation was relatively small.

Several researchers from the 1930s have some claim to have invented this form of low-drag truncated rear, but the configuration is popularly known as the Kamm-back after W. Kamm, who developed it via a series of prototypes (Fig 4.16). Wind-tunnel measurements on one of the surviving vehicles (see Hucho[11]) in 1979 yielded a drag coefficient of 0.37, which was very low for its day. The resulting shape was unfortunately considered ugly at the time in comparison with conventional designs. The current approaches to rear-end treatment are described later in this chapter.

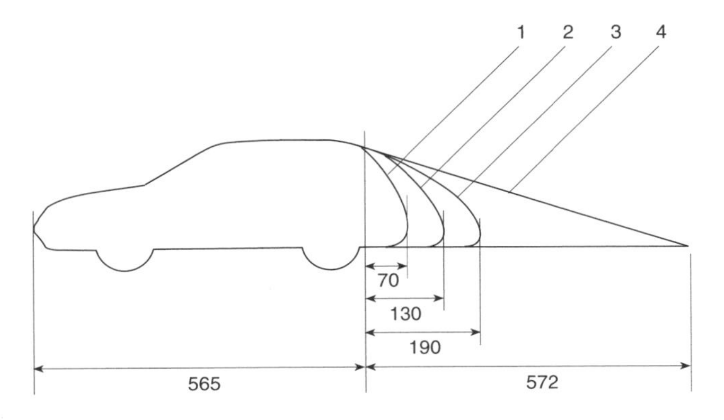

Configuration number	C_D
1	0.30
2	0.23
3	0.21
4	0.12

Fig. 4.14 Some results from W. E. Lay's study of basic car shapes, showing the effect of reducing the tail length.

Fig. 4.15 The stylish curved rear end, popular in 1930s American cars, produced high drag and rear-end lift.

Fig. 4.16 One of the original K-form or Kamm-back vehicles designed in the 1930s by W. Kamm, with a C_D of 0.37.
(*Photograph by courtesy of Volkswagen AG, and Deutsche Forschungsanstalt für Luft- und Raumfahrt*)

THE CONSERVATIVE APPROACH - EVOLUTION

Despite the many attempts at producing low-drag streamlined vehicles, most production cars throughout the world use shapes that have evolved gradually from the box-like forms of horse-drawn carriages. Figure 4.17 illustrates the basic forms that have emerged. The three-box version is still the most popular, although in Europe, the two-box hatchback or fastback now comes a close second. A more recent trend has been the evolution of single-box family cars such as the Renault Espace shown in Fig. 4.18, and the Renault Twingo shown in 4.19. The latter

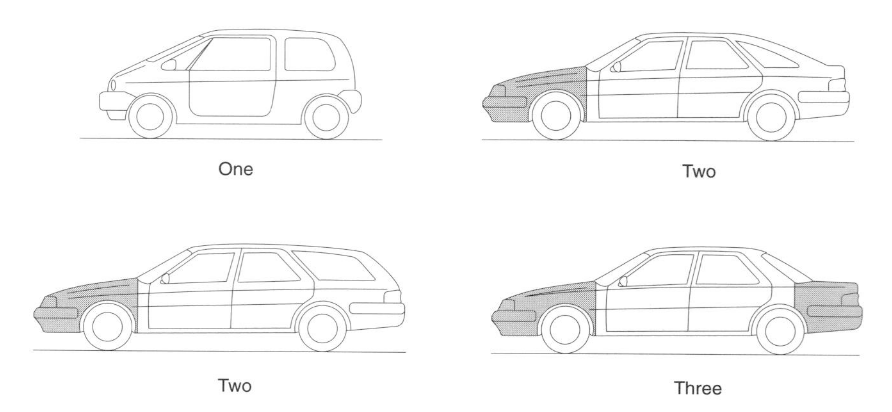

Fig. 4.17 The basic one-, two-, and three-box forms.

Fig. 4.18 The Renault Espace
Typical of the newer generation of single-box forms based on commercial vehicles.

Fig. 4.19 The Renault Twingo
An example of the application of the single-box form to a small car.

example is interesting as it is a small vehicle designed specifically as a car, and not derived from a van. Public acceptance of the single-box form opens up interesting possibilities for the future, such as a re-evaluation of the radical teardrop-based approach pioneered by the Hansen and Schlör concept vehicle (Fig. 4.5).

Some attempts at streamlining the basic three-box form were made from the 1930s to the 1950s, particularly in the USA, but these were generally more a matter of fashion than a serious desire to lower the drag resistance. The shapes simply reflected the curved forms of the art-deco style, and the 'modernist' preoccupation with things that looked fast and sleek. The major impetus to serious aerodynamic

Fig. 4.20 A typical box-like vehicle of the pre-oil crisis period.

drag reduction came in 1973 when, as described in Chapter 3, the OPEC cartel forced a drastic rise in the price of fuel. The angular forms popular at that time (Fig. 4.20) were particularly bad in terms of aerodynamic drag because of the separations produced at the sharp edges. The motor manufacturers' short-term approach to aerodynamic improvements consisted mainly of introducing small deviations from the shape of their previous models, with no radical change of style. Hucho[11] cites a number of examples of how drag coefficient reductions of 10 per cent or more were obtained by relatively small modifications to the basic shape of German cars of that period.

A breakthrough in Europe occurred in the early 1980s when the multinational manufacturers finally started to introduced genuine low-drag shapes based on wind-tunnel research. Cars such as the Ford Sierra (Fig. 4.1) with a C_D of around 0.34 (there are differences between versions) looked quite different from vehicles of the previous generation, although they still essentially conformed to the two- or three-box forms.

THE FLOW AROUND A SIMPLE THREE-BOX VEHICLE SHAPE

To explain how low-drag shapes are evolved from the traditional basic forms, it will be helpful to start by describing the flow past a simple three-box shape. Figure 4.21 illustrates some of the undesirable flow features of the basic unrefined three-box shape. The lower part of Fig. 4.21 shows the flow pattern on the centre-line. Separation occurs from the top of the near vertical radiator grill at (a). After reattachment at (b) some way back on the engine cover, the flow separates again at (c) in front of the windscreen. After reattaching on the glass at (d) it again separates at the top corner (e). Reattachment follows on the top at (f) followed by a final separation at the end of the roofline (g). Partial and unsteady reattachment may

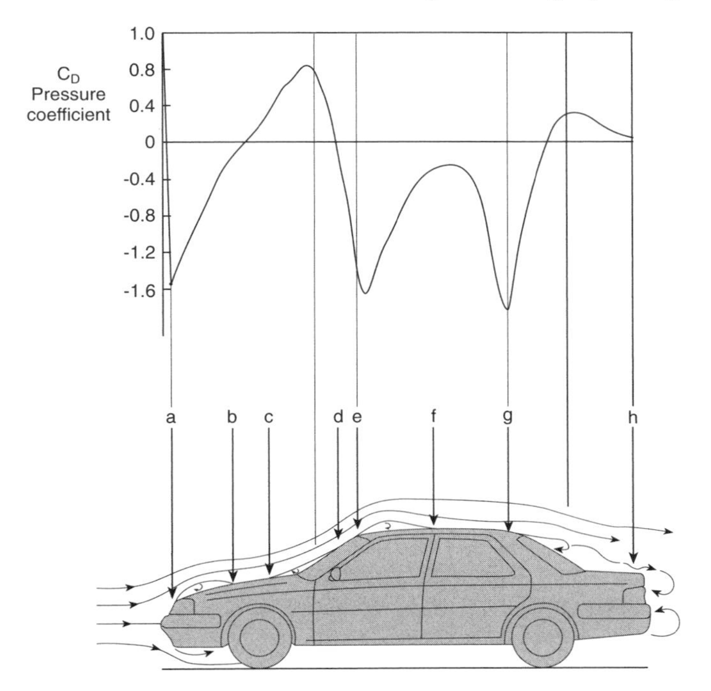

Fig. 4.21 Centre-line pressure distribution and flow features of a simple three-box form vehicle.

occur again at (h). The flow separations are caused by strongly adverse pressure gradients (pressure increasing rapidly in the direction of flow). From the pressure distribution, it may be seen that the separation positions correspond with regions of strongly adverse pressure gradient.

Other drag-producing aspects of the flow around the traditional three-box vehicle are the conical vortices which are generated by separations from the A-posts at the front, and at the rear, as described in Chapter 2. Removing such undesirable features is not particularly difficult, and in the sections below, we describe the approaches required.

IMPROVING THE FRONT END

The most obvious possibility for improvement of the basic three-box shape lies in avoiding the flat front and sharp corner above the radiator (position (a) in Fig. 4.21), which produced strong separations on older vehicles. Hucho[11] describes drag reductions of up to 14 per cent produced by modest rounding and lowering of this corner. Much more significant improvements however are obtained when the front end is made as a smooth continuous curve originating from the line of the front bumper. This form was pioneered by Citroën in the DS design shown in Fig. 4.7 and is continued in the more recent XM (Fig. 4.6).

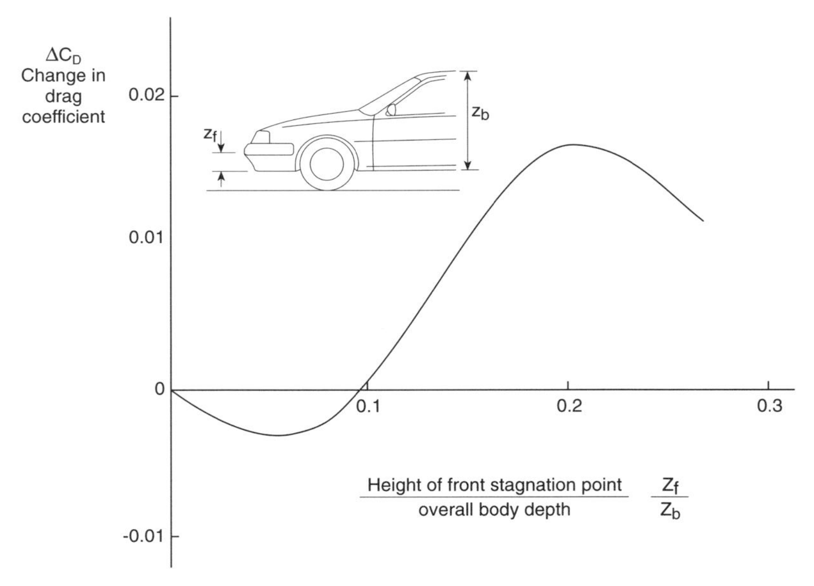

Fig. 4.22 The effect of front-end height.
(*Based on data in Buchheim et al.*[2])

Figure 4.22, which is based on data presented by Buchheim *et al.*,[2] shows how the drag coefficient is strongly influenced by the height of the leading edge, which effectively defines the stagnation line location. A low position is clearly required. This work indicates that there is an optimal minimum, but it is not safe to generalize too much from one case. It should be noted that a low front stagnation line also has the advantage that it tends to reduce the lift at the front.

Using the techniques of computational fluid dynamics described in the final chapter, it is possible to design a front-end profile that provides a steadily decreasing pressure almost up to the front screen. This not only inhibits separation, but produces a low rate of boundary layer growth, and provides the possibility of producing a significant area of low-drag laminar boundary layer.

THE EFFECT OF SCREEN RAKE ANGLE

With the classic three- and two-box shapes, an area of high pressure occurs just upstream of the front screen, often with a separation bubble of recirculating air at the base of the screen (see Fig. 4.21), and this is a source of drag. The size and influence of the high-pressure area depend on the front screen rake angle; making the screen more raked tends to reduce the pressure at the base of the screen, and to lower the drag. Much of the drag reduction associated with increasing the screen rake angle, however, is due to the fact that a high rake angle enables the curvature of the corner between the screen and the roofline to be reduced, thereby helping to maintain attached flow. Figure 4.23 shows an example of a well-curved roofline, a style that might have been considered unacceptably 'dumpy' a few years previously. There is a practical constraint on the screen rake angle that can be used, because high angles produce problems of solar heating and internal reflections from the glass. Sports cars such as the Jaguar shown in Fig. 4.24 often have a very high rake angle, but sports cars are not expected to be totally practical.

Fig. 4.23 The Vauxhall Corsa
Showing a well-curved roofline and a tapered and rounded truncated rear.

Fig. 4.24 Sports cars often have a high front screen rake angle, but this can produce problems of solar heating and internal reflections from the glass.

There have been many wind-tunnel studies of simplified vehicle shapes, aimed at evaluating the effects of the principal geometric parameters such as front and rear screen rake angles. Scibor-Rylski[24] shows how separation and reattachment points are affected by screen rake angle for a simple block (straight-sided) model. It is however not possible to totally isolate effects in this way for realistic vehicle shapes. The flow over each part of the body influences the flow over every other part to some extent, and furthermore, the flow can be greatly influenced by three-dimensional effects. For example, curving the windscreen in plan view modifies the flow patterns considerably, as it facilitates a cross-flow, which reduces the extent and intensity of high pressure.

In practice, the small area of high pressure just in front of the screen is useful, as it provides a good location for a fresh-air intake. Reattachment of the flow normally occurs between about a quarter and a third of the way up the screen on modern designs, which typically have rake angles of around 55–60° and significant curvature.

In the single-box design of the Renault Espace (Fig. 4.18) and Twingo (Fig. 4.19), the windscreen forms part of a continuous curve of the centre-line profile at the front, and the high-pressure region at the base of the windscreen is eliminated. With this form of vehicle it should be possible to provide a favourable or only mildly unfavourable pressure gradient up to the point of maximum vehicle height. It may also be noted that single-box shapes can be readily derived from the starting point of an ideal teardrop with truncated rear.

THE EFFECT OF THREE-DIMENSIONAL FLOWS

The discussion above has concentrated on the flow along the centre-line profile. It must be remembered, however, that a road vehicle is a highly three-dimensional shape, and the air flow generally has a sideways component. The pressure on the sides of the vehicle will mostly be around or little below the ambient atmospheric value, whereas the pressure on the upper surfaces is generally lower, and there will therefore be an upward component of surface flow. As described above, however, in the vicinity of the base of the windscreen, the pressure is high, so in this region, the flow will tend to spill over from the top surface and down the sides. This effect may be seen in the wool-tuft visualization picture shown in Fig. 4.25.

Separation will occur if the resultant flow encounters a sharp edge, so it is important to have smoothly rounded edges everywhere, and not just in the centre-line profile. A strong outward cross-flow can occur towards the edges of the windscreen, tending to produce separated vortices around the A-posts, as described previously. These vortices are sources of both drag and noise. To inhibit their formation, it is necessary to ensure a smooth curve on the A-post, combined with some curvature of the screen itself. Figure 4.26 shows the method of windscreen attachment and A-post detail on a 1980s design compared with that of an older vehicle. The elimination of pronounced rainwater gutters and the use of a bonded windscreen as opposed to one held in by a rubber moulding help to maintain

Fig. 4.25 A small region of down-flow produced by the high-pressure area in front of the windscreen disturbs the flow along the side, as shown by the wool tufts on this wind-tunnel model.

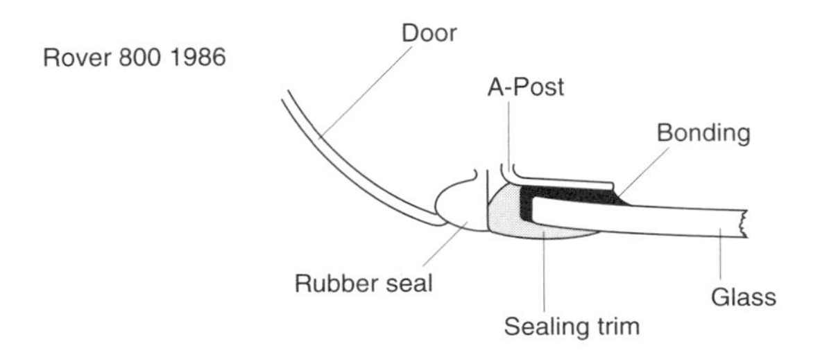

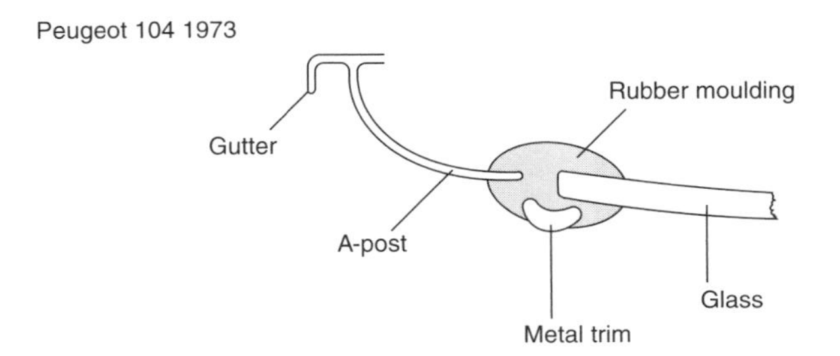

Fig. 4.26 Detailing on the A-post showing how the profile is smoother and more continuous on more recent vehicles.

attached flow. Details of the design development of the A-post of the VW Scirocco are given by Hucho.[11] A drag coefficient of 0.37 was found for the best case tested, as compared with 0.41 for the worst. These figures clearly demonstrate the value of attention to detail in the A-post area. Similar separations can occur at any sharp junction between horizontal and vertical surfaces.

THE REAR END

There are three common rear-end forms for private cars as illustrated in Fig. 4.27. They are usually known as the notchback, the hatchback (or fastback), and the squareback. The hatchback looks to be the most promising from the aerodynamic viewpoint, but this can be deceptive. Behind the rear screen there is a region of low pressure, and conical vortices tend to form around the rear corners, as illustrated in Fig. 4.28. These are in part the origin of strong trailing vortices. The vortices draw in the boundary layer from the rear screen or backlight, so that attached flow can be maintained on the screen even at rake angles of 30°. The problem is that by maintaining flow attachment, the air is pulled down strongly at the rear. The reaction to this change of flow momentum is the production of a force on the car which has both lift and drag components. The fact that strong trailing vortices are produced shows that lift is being produced, and clearly, trailing vortex drag will be generated. The styling model shown in Fig. 4.29 was found to have a drag coefficient of 0.55 despite its superficially streamlined appearance. The flow remained attached to the rear screen despite the large rake angle.

Figure 4.30 shows the variation of C_D and C_L with rear screen rake angle based on data from various sources including the author's own. At very small rake angles the shape is not really a hatchback at all, but more of a tapered squareback. The drag coefficient initially falls with increasing rake angle, because raking the back down increases the amount of taper, and the normal pressure (form) drag therefore tends to decreases. At about 10°, however, the drag starts to rise due to the formation of strong conical vortices. The experimental results show a peak in the drag coefficient at a rake angle of around 30°. At angles greater than 30°, stable

Fig. 4.27 Rear end forms: notchback (left); hatchback (centre); squareback (right).

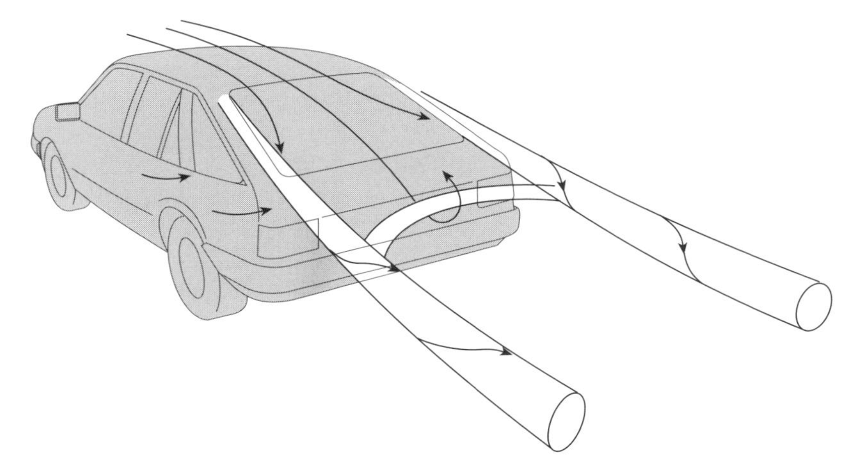

Fig. 4.28 Conical vortices are generated by flow separation from the rear corners on a hatchback, and become the origin of strong trailing vortices.

Fig. 4.29 A hatchback styling model which gave a C_D of 0.55 despite a superficially streamlined appearance. Strong conical vortices at the rear maintained attached flow over the rear screen with attendant lift and drag penalties.
(*Photograph courtesy of Terry Newman*)

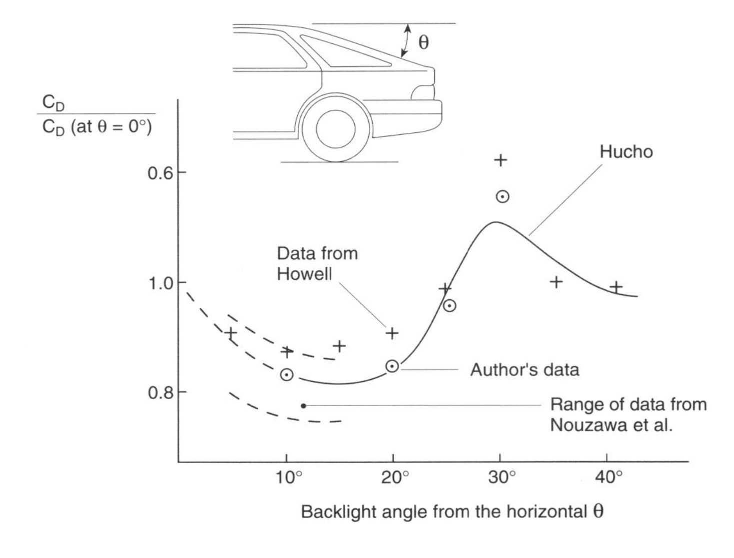

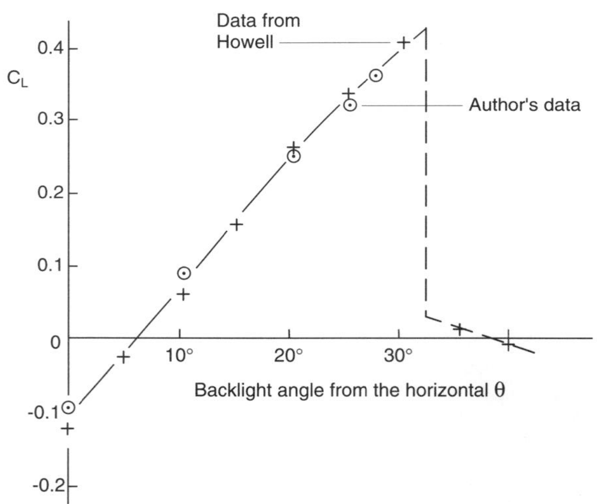

Fig. 4.30 The variation of C_L and C_D with backlight angle, based on data from various sources.

vortices cannot form, and the flow separates. Thus, at high rake angles, the rear end begins to behave as a squareback; the trailing vortex drag falls, but there is an increase in normal pressure drag. There is some scatter in the drag data from the various authors, but this is due to differences in the vehicles used. The peak at 30° and minimum at around 10° are evident in all cases.

The lift increases nearly linearly with backlight angle up to about 30°, where the flow then separates at the end of the roofline, producing a stall or sudden fall-off

Fig. 4.31 The VW Polo
Showing rear-end taper in plan and side views for low drag. The squareback form is unusual as a basic body style for a small popular car.

in lift. The high level of agreement between the author's data and that from Howell[10] must be largely a matter of chance, as the models used were quite different. Howell used a simplified shape similar to that of Ahmed[1] (see Chapter 12), whereas the author's data were obtained from a modified model of an old Austin 1800.

The high drag associated with rake angles around 30° leads to the rather surprising result that the squareback estate-car (station-wagon) version of some vehicles has been found to produce less drag than the apparently more streamlined hatchback. This behaviour also explains why it is sometimes better to cut the rear off as in the Vauxhall Corsa (Fig. 4.23) rather than curve it downward at the rear as in the case of the VW Beetle. The more recent VW approach applied to the Polo is shown in Fig. 4.31; the form is almost that of a squareback estate. These relatively modern designs follow the basic principle of the Kamm-back, and show good tapering and rounding at the rear.

THE NOTCHBACK REAR

The flow behind notchback vehicles has been extensively studied, and has been found to contain a complex arrangement of standing and trailing vortices (see Nouzawa *et al.*[23] and Ahmed.[1] An important factor is the angle θ_{eff} made by a line joining the rear end of the roofline to the tip of the boot (trunk). As shown in

Fig. 4.32, the variation of drag with this angle shows similarities to the drag variation with rear screen slope angle for a hatchback. The importance of the angle θ_{eff} ties in with the finding that raising and/or lengthening the boot (trunk) generally reduces the drag. Clearly if the boot line is raised, the angle θ_{eff} will decrease. The increase in the boot height on the revised Rover 800 series produced a useful drag reduction. Improvements to the drag by raising and lengthening the boot of the Audi 100 are described by Buchheim *et al.*[3]

The flow in the profile centre-line plane generally shows a pair of vortices rotating in opposite senses at the rear, as illustrated in Fig. 4.33. On lifting

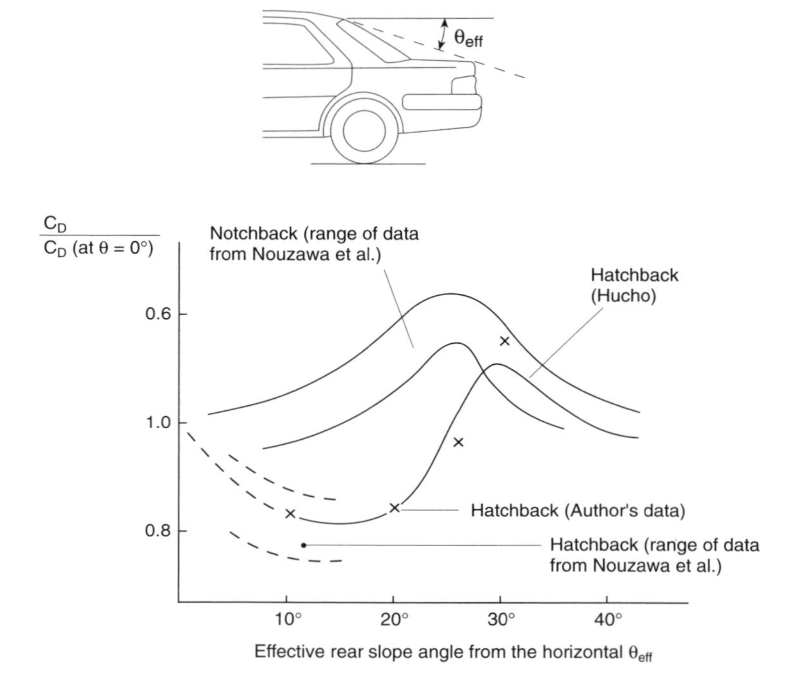

Fig. 4.32 The influence of the effective rear slope angle θ_{eff} on C_D for a notchback vehicle.

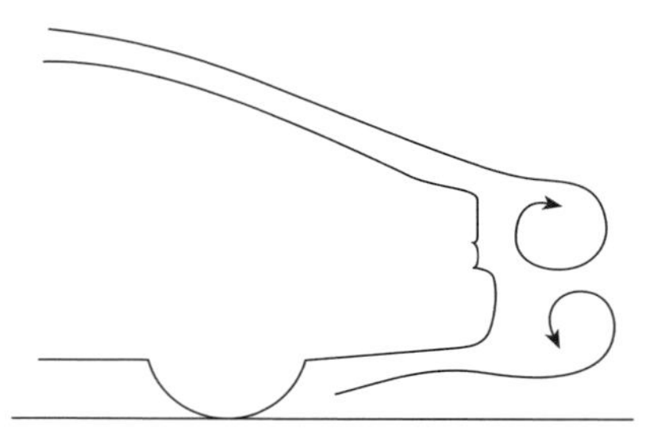

Fig. 4.33 Standing vortices in the wake on the centre line.

hatchback shapes, the upper vortex dominates and links into the trailing vortices. In the squareback configuration, the two vortices are generally nearly equal, the difference between them making a weak addition to any trailing vortex.

ROOFLINE CURVATURE

The general principle of maintaining favourable or mildly unfavourable pressure gradients by use of smooth contours applies to the shape of the main roof area. Buchheim *et al.*[2] show that increasing the curvature of the roofline reduces the drag coefficient. However, as Buchheim *et al.* also show, if the increase in curvature is simply produced by bulging the roofline up, then the resulting increase in projected frontal area may more than offset the reduction in drag coefficient, and the overall $C_D A$ value may increase. The alternative of bringing the roofline down at the front and rear is restricted by the requirements of driver sightlines.

PLAN-VIEW SHAPING

Because a car is a three-dimensional object, it is better to think in terms of sheets of flow, stream-surfaces rather than streamlines. This is however conceptually difficult, and it is easier to start by considering the flow around the centre-line profile and plan views separately.

Rounding the front-end corners in plan will undoubtedly reduce drag, and popular cars of the 1990s have a far more rounded frontal aspect from all views. Rounding should be applied to all forward-facing elements such as the A-posts.

Increasing the plan-view curvature will also normally reduce the drag coefficient, but Buchheim *et al.*[2] show that as with roofline camber, the drag factor $C_D A$ will rise if the increase in curvature is obtained by widening the vehicle.

Tapering the rear in plan view (boat-tailing) can produce a significant reduction in drag, as it provides the required region of pressure recovery via a mildly unfavourable pressure gradient. Hucho *et al.*[12] show a drag coefficient reduction from 0.42 to 0.37 obtained by modest boat-tailing aft of the rear wheel arch.

Plan-view tapering is symmetrical, unlike the usual method of side-profile tapering, which involves curving the line down to the rear, as typified by the VW Beetle. Unlike the side-profile tapering, therefore, planform tapering does not produce trailing vortex drag. Many cars such as the VW Polo (Fig. 4.31) now display a noticeable amount of rear-end taper.

UNDERBODY PROFILING

Amongst the rash of low-drag concept cars that appeared in the 1980s was the BL Technology ECV3. During the early stages of the development, it was found that it was impossible to reduce the drag coefficient to 0.3 whilst retaining the rough and disorderly underside that was conventional for production vehicles of the time. When the floor-pan was redesigned to give a smoothed continuous and largely unbroken surface, a final C_D value of 0.25 was obtained. This clearly demonstrates the importance of the underbody design.

Traditionally, apart from the various protrusions, the overall profile of the underside has been more or less flat, but as described earlier, for a body in moderate proximity to the ground, the ideal shape would have some curvature on the underside. Figure 4.34, which is based on data from Buchheim *et al.*[2] and Howell,[10] shows the advantages of providing a curved diffusing (pressure rising towards the rear) underside at the rear. The upsweep of the underside increases the rearward taper thus tending to reduce the form drag. It also lowers the lift-dependent trailing vortex drag by reducing both the camber and the effective angle of attack of the vehicle. The rear-end diffuser produces a reduced pressure under the vehicle, as described in more detail in Chapter 6, and this consequently produces a reduction in lift and trailing vortex drag. A smoothly curved underside profile will also provide a worthwhile reduction in surface friction drag.

From Fig. 4.34 it may be seen that the variation of drag with diffuser length and angle is complex, and it is possible to end up with an increase in drag. Howell's work[10] also shows that there is a complicated relationship between the diffuser and backlight geometries on a hatchback. The upper and lower surfaces cannot therefore be optimized independently, and the rear end has to be treated as a single three-dimensional shape. Clearly, changes of either the upper side or underside slope angle will affect both the camber and the effective attitude of the car.

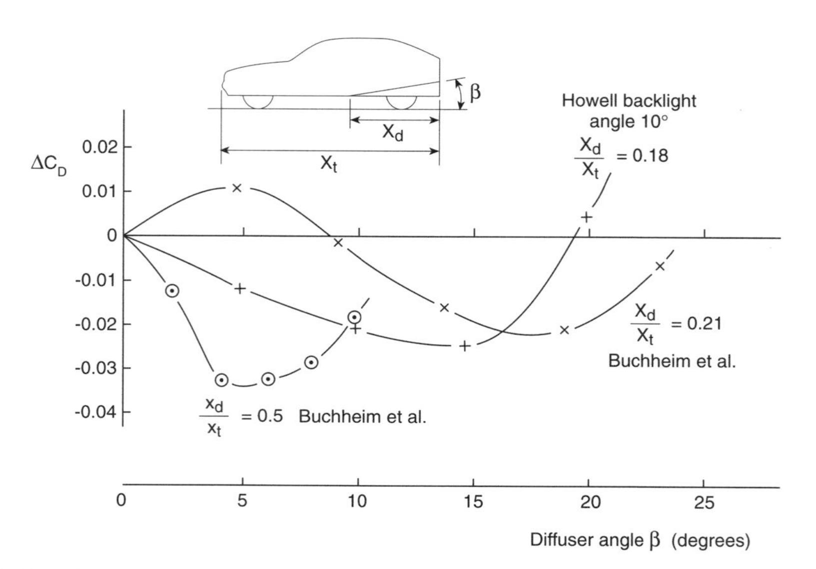

Fig. 4.34 The effect of underbody tapering at the rear to produce a diffusing area. (*Based on data presented in Buchheim et al.*[2] *and Howell*[10])

In the past, providing a smooth underside was not considered to be worthy of the expense incurred, but it is now recognized that the design of the underside represents one of the most fruitful areas for drag reduction (see Carr[4]). Lightweight plastic fairings are commonly used in the front portion of the engine bay, and some manufacturers have started to produce floor-pans that accommodate details such as the exhaust system conformally into recesses in an otherwise smooth continuous surface. Items such as the engine sump which require a cooling airflow are set flush with the surface in well-fitting cutouts. Once a floor-pan of this type has been designed, it becomes a relatively straightforward matter to adapt it for future models.

DRAG CHARACTERISTICS OF BASIC DESIGN FEATURES

An insight into the influence of various styling features of the pre-oil crisis cars was provided by White[25] of MIRA, who made extensive wind-tunnel tests on a large number of vehicles. From his results he was able to assign an empirical drag 'rating' value to the design features of nine identified zones such as front outline plan, and front outline elevation. For example, a front plan of semicircular form was given a rating of 1, whereas a squared constant-width front scored 6. By adding up the rating values for each of the nine zones, a final score was arrived at that gave a corresponding C_D value. The method produced surprisingly good results at the time, but it is largely redundant now, as many of the styling features have disappeared, and there is a great deal of similarity between the basic body shapes. Improvements in the drag of current cars are now mostly obtained by careful attention to detailed design rather than by producing new shapes.

AIR DAMS

Air dams were initially introduced as a bolt-on extra which was used to improve the lift and drag characteristics of an existing design of vehicle. The main intention of the air dam was to reduce the pressure on the underside of the vehicle, thereby reducing the lift and the attendant trailing vortex drag. As illustrated in Fig. 4.35, the rear of the vehicle is at a relatively low pressure. By restricting the gap between the road and the car at the front by means of a dam, the air flow speed is increased locally, after which it gradually slows down again as the area increases towards the rear. Following the Bernoulli relationship, this means that the pressure under the front part of the vehicle must be lower than at the rear (where it is already lower than ambient). The low pressure on the underside results in a negative contribution to lift, particularly at the front end. Towards the rear, air flowing in from the sides tends to reduce the suction effect. Air dams are therefore used to control both the drag and the lift distribution. The benefits of reduced front-end lift are relatively slight for normal road-going cars, but for high-speed racing vehicles they are of major importance, as described in Chapter 6.

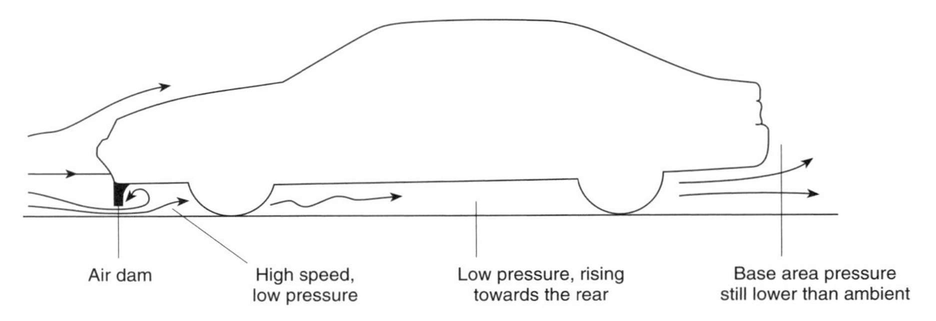

Fig. 4.35 The effect of an air dam
The low pressure under the car reduces lift, and can reduce lift-related trailing vortex drag.

It might be thought that adding a large bluff element to the front of the car would increase the pressure drag, but as long as smooth attached flow is maintained, this need not be so; the dam merely lowers the position of the front stagnation line which, as shown previously, is generally advantageous. Some additional pressure drag is incurred if the rear face of the dam is exposed to the low pressure, and it is better to incorporate the dam effect into the basic design, with the underside joining smoothly to the bottom of the front panel.

Although air dams do work if properly designed on the basis of wind-tunnel testing, their popularity was partly a matter of fashion, and proprietary add-ons were not always effective (see McCarthy[20]). Nowadays, the same effect is achieved by means of an appropriate shape of front end, curved down to the front to produce a low stagnation line. In general, it is better to design a smooth well-contoured underside with an overall vehicle shape that produces no lift. Down force on passenger cars carries a drag penalty with little advantage in terms of roadholding at legally permitted speeds.

REAR-END SPOILERS

Rear-end spoilers are normally in the form of an upturn at the end of the boot (trunk) lid. Their purpose is to improve roadholding by reducing rear-end lift. The effect is similar to that of raising the boot height described previously. The line between the end of the roof and the tip of the spoiler defines the effective rear slope angle. Below the critical angle, any reduction in angle tends to reduce the drag. Reducing the lift will lower the trailing vortex drag, but once again, this may be offset by an increase in normal pressure drag. The roadholding improvements are only likely to be noticeable on the racetrack. Road tests described in Carr[4] show that spoilers like air dams can be effective in reducing both lift and drag, as long as they are based on experiments, and not just intuition. In some high-performance vehicles, very large spoiler-like elements attached to the rear screen may be employed, but these are more in the nature of an inverted wing intended to produce down force at the rear to improve tyre adhesion. The use of inverted wings is described in Chapter 6.

LOW-DRAG CAR DESIGNS

In the 1980s, several low-drag concept cars were developed, either by private enterprise, as in the USA, or with government backing, as in France and Germany. Table 4.1 lists the salient features of some of them. Apart from the Probe IV and V designs, the cars were near-practical road vehicles, in some cases based on existing models. Of particular interest was the Probe III, which was developed by Ford of Europe, and bears a considerable resemblance to the Ford Sierra. Its low drag

Table 4.1 Some low-drag experimental cars of the early 1980s developed in response to the post-oil crisis demands for low-drag economical vehicles.

Car	C_D	Comments
Ford (USA)		
Probe I (1979)	0.25	Sports coupé roughly based on dimensions of a Mustang
Probe II (1980)		Family saloon based on same principles
Ford of Europe		
Probe III (1981)	0.22	Similar in appearance to the Sierra
Ford (USA)		
Probe IV (1982)	0.152	Very much a concept car with several rather impractical features such as flexible front wheel covers
Ford (USA)		
Probe V (1985)	0.137	Another pure concept car
UNI-CAR (1981)	0.228	A joint German government industry project with support from academic institutions
BL ECV_3	0.25	A reasonably practical vehicle
VW Auto 2000 (1981)	0.25	A practical vehicle resembling the subsequent Polo
Peugeot VERA 01 (1980)	0.306	Based on a Peugeot 305, which in its original form had a C_D of 0.44
Peugeot VERA PLUS (1983)	0.22	A major development with smooth underpan, highly raked screen, covered wheels etc.
Renault VESTA	0.22	A fairly practical vehicle

coefficient of 0.22 compared with 0.34 for the Sierra was to a significant extent attributable to underbody and cooling system refinements.

It should be remembered that the drag of the final engineered vehicle will depend strongly on the details. Figure 4.36 shows how remarkably smooth contours can be obtained on standard production vehicles by using such features as flush-mounted glazing, and by well-developed production design.

As has been described, it is not too difficult to produce a basic shape (with wheels) that displays a drag coefficient of 0.15; the difficulty lies in translating the basic design into a practical vehicle that can be cheaply mass-produced. This final stage is where the battle is won or lost. Typically the drag coefficient will double during final detailing. The wheel trims alone can change the drag coefficient by 0.02. With new production techniques and careful attention to detail, including underbody flows, it should be possible to achieve C_D values of around 0.2–0.25 without resorting to extreme forms. This can only be achieved, however, if the stylist has a proper understanding of the principles of vehicle aerodynamics, and is prepared to restrain the urge to express his personality in a visual form that is at odds with the requirements of aerodynamics. Breaking through the C_D of 0.2 barrier is a challenge that has been addressed in the concept cars, but for practical designs, this probably represents an area of diminishing returns. It is easier to make fuel savings by other means.

Fig. 4.36 Flush glass mounting and careful attention to panel joint lines produce a good continuously curved smooth surface.

CARAVAN TOWING

Private cars have to fulfil a number of roles, and one of these is the occasional requirement to tow a caravan. In some cases, the summer holiday journey may represent quite a significant part of the annual mileage, so drag reduction of car/caravan combinations is of some importance.

Attaching a caravan behind a car produces a large increase in aerodynamic drag. Hands and Zdravkovich[7] and Garry[5] show drag coefficients for car–caravan combinations (based on the projected frontal area of the caravan) in the order of 0.9 for zero yaw angle. Since the caravan typically has nearly twice the frontal area of the car, the C_D value based on the car's frontal area can be up to 1.8. The caravans studied by Garry were short dumpy units of around 4.25 m (14 ft) length. A C_D value of 0.55 is quoted by Götz[6] for a more luxurious combination of a large car and relatively long caravan, and a value as low as 0.35 for a combination of a well-streamlined caravan and a low-drag (for 1981) Mercedes towing car.

The reason for the high drag of most car–caravan combinations is evident in the flow visualization pictures of the above two references. As illustrated in Fig. 4.37, the flow meeting the stagnation line along the front of the caravan has merely been deflected by the car, and still has the full free-stream energy. The drag of the combination is therefore likely to be similar to the sum of the drags of the two vehicles measured independently. Modern cars tend to be worse in this respect than older vehicles, which often produced a wide separated wake in which the caravan was immersed.

There are two obvious methods of approaching the problem. The first is to streamline the caravan by rounding its front corners. Garry's results show 17 per cent less drag for a streamlined caravan compared with a conventional sharp-edged model. The second method is to mount a deflector plate or fairing on the roof of the car. The purpose of the deflector is to guide the shear layer from the top of the car so that it reattaches on the top of the caravan, as shown in Fig. 4.38. The front surface of the caravan will then be in the low-energy low-pressure wake of the car.

The effectiveness of roof-mounted devices was tested by Garry[5] both in the wind-tunnel and by track testing full-size vehicles. A moulded integrated fairing proved to be more effective than simple plate-like devices, particularly when the

Fig. 4.37 Flow patterns, baseline car/caravan combination. (*Photograph courtesy of K. P. Garry*[5])

Fig. 4.38 Flow patterns, car with rooftop deflector and caravan. (*Photograph ourtesy of K. P. Garry*[5])

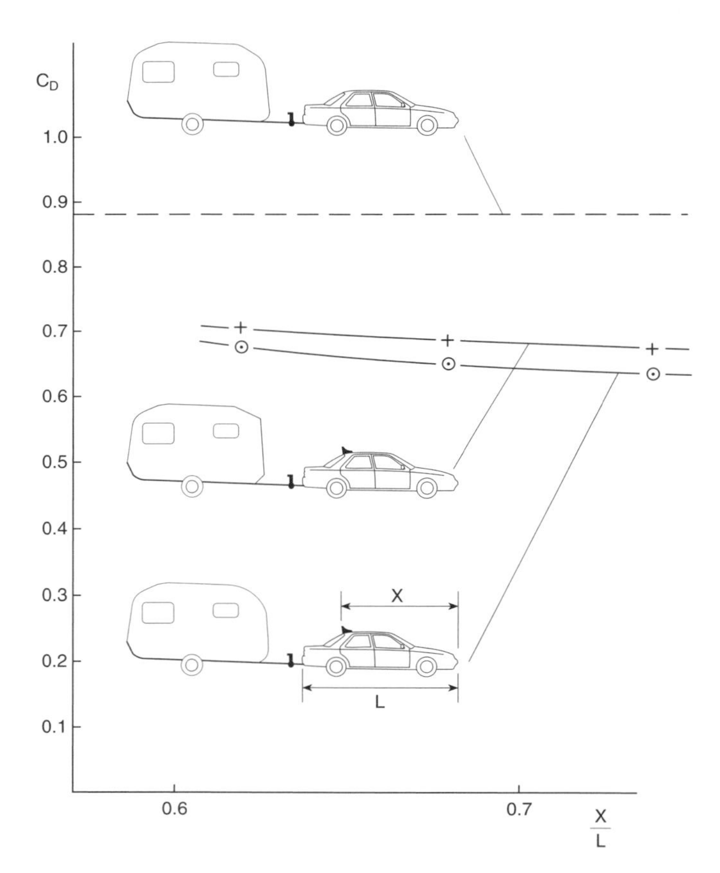

Fig. 4.39 The effect on drag of the longitudinal positioning of a moulded rooftop deflector for a standard caravan, and for a modified caravan with a streamlined leading face. (*After Garry*[5])

vehicles were yawed. It was also found that the best position for the fairing was as far aft as possible. As a consequence, the squareback estate car produced the lowest drag coefficient. Figure 4.39, which is based on Garry's data, shows the variation of C_D with the longitudinal positioning of a moulded rooftop deflector for a saloon car with a standard caravan, and also with a modified caravan with a streamlined leading face. It will be seen that the deflector produces a significant drag reduction, and that streamlining the leading face further reduces the drag. Full-scale track tests gave a best fuel saving of 16.6 per cent for a combination of an estate car with a moulded roof fairing, and a standard Sprite caravan.

Although roof-mounted devices produce significant improvements at zero yaw, they are less effective at large yaw angles. Garry's results (Fig. 4.40) show that the plate-like deflector is worse in this respect than the moulded deflector, and that all the devices become ineffective at high yaw angles. The results of Götz show that a deflector can produce a large increase in drag at high yaw, and that there are considerable advantages in adopting a streamlined caravan shape. Car–caravan combinations cover a wide variety of configurations, and it is only possible to draw rather general conclusions from the published data.

Significant fuel savings can undoubtedly be made by fitting a deflector, but some of the benefit can be lost if after parking and unhitching the caravan, the car is driven around for the rest of the holiday with the deflector in place. The simple deflector plate, though less effective than the fairing, particularly in cross-winds, has the advantage that it can be folded flat when not required.

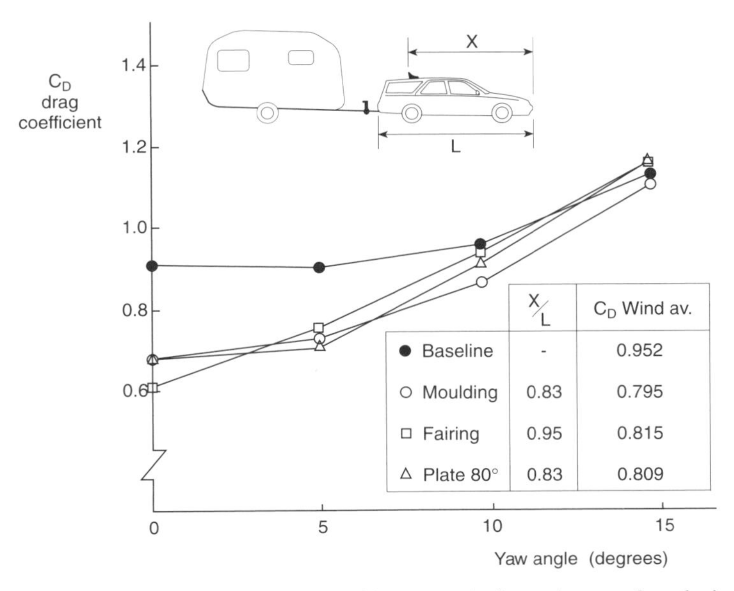

Fig. 4.40 The variation of drag coefficient with yaw angle for various rooftop devices located at their optimum position: estate car with a standard Sprite caravan. (*After Garry*[5])

One final method of improving the drag of a car–caravan combination is to fit a long streamlined extra-baggage fairing on the roof of the towing vehicle. Single-box vehicles of the Renault Espace type would appear to be ideal for this purpose.

DESIGN PRINCIPLES FOR LOW DRAG

In this chapter there has been some examination of design details in isolation, but in designing a low-drag vehicle, it is important, as far as possible, to treat it as an integrated whole. The essential features for low-drag design are as follows.

1. Smooth unbroken contours with favourable pressure gradients as far back as practical should be used.
2. Strongly unfavourable pressure gradients at the rear should be avoided; some taper and rear-end rounding should be used.
3. The form should produce negligible lift in ground proximity.
4. If a hatchback configuration is required, the backlight angle should not be in the region of 30°, and if a notchback is is to be used, the effective slope angle θ_{eff} should not be in this region.
5. The underbody should be as smooth and continuous as possible, and should sweep up slightly at the rear.
6. There should be no sharp angles (except where this is necessary to avoid cross-wind instability: see Chapter 10).
7. The front end should start at a low stagnation line, and curve up in a continuous line.
8. The front screen should be raked as much as is practical.
9. All body panel lines should have a minimal gap.
10. Glazing should be flush with the surface as much as possible.
11. All details such as door handles should be smoothly integrated within the contours.
12. Excrescences should be avoided as far as possible; windscreen wipers should park out of the airflow.
13. Minor items such as wheel trims and wing mirrors should be optimized using wind-tunnel testing.
14. The cooling system needs to be designed for low drag (see Chapter 7).

Many contemporary vehicles show some or most of these features, and it is an interesting pastime to see how many can be spotted on individual models. Fortunately there is no single ideal shape. Acceptably low drag can be achieved with a wide variety of forms, so there is little danger of variety totally disappearing in automotive shapes.

This chapter has dealt principally with the problems of designing domestic cars for low drag; later chapters examine other important aspects of the aerodynamic design, such as heating, cooling, ventilation and noise.

REFERENCES

1. Ahmed S. R., An experimental study of the wake structure of typical automobile shapes, *Proc. 4th Colloquium on Industrial Aerodynamics*, Fachhochschule Aachen, June 1980.
2. Buchheim R., Deutenbach K. R. and Lucköff H. J., Necessity and premises for reducing the aerodynamic drag of future passenger cars, SAE paper 810185, Detroit, 1981.
3. Buchheim R., Leie B. and Lückoff H. J., Der neue Audi 100 – Ein Beispiel für Konsequente Aerodynamische Personenwagen-Entwicklung, ATZ 85, 1983, pp. 419–25.
4. Carr G. W., Aerodynamic effects of underbody details on a typical modern car model, MIRA rep. 1965/7.
5. Garry K. P., Wind-tunnel investigation into the effectiveness of drag reducing devices for car-caravan combinations, *Proc. Colloquium, Designing with the Wind*, CSTB Nantes, 15–19 June 1981, pp. XI-1-1–15.
6. Götz H., Examples of practical aerodynamics in passenger cars, caravans and buses, paper XI-6, *Proc. Colloquium, Designing with the Wind*, CSTB Nantes, 15–19 June 1981, pp. XI-6-1–20.
7. Hands S. J. and Zdravkovich M. M., Drag reduction for a passenger car towing a caravan, *Proc. 4th Colloquium on Industrial Aerodynamics*, Fachhochschule Aachen, 18–20 June, 1980, pp. 223–32.
8. Hansen M. and Schlör K., *Der AVA-Versuchwagen AVA-Bericht 43*, W 26, Göttingen, 1943.
9. Howard G., *Automobile Aerodynamics*, Osprey, London, 1986.
10. Howell J., The influence of ground simulation on the aerodynamics of simple car shapes with an underfloor diffuser, *Proc. RAeS Conference on Vehicle Aerodynamics*, Loughborough, July 1994, paper 36, pp. 36.1–11.
11. Hucho W. H., *Aerodynamics of Road Vehicles*, Butterworth, London, 1987.
12. Hucho W. H., Janssen L. J. and Emmelman H. J., The optimization of body details – a method for reducing the aerodynamic drag of road vehicles, SAE paper 760 185, Detroit, 1976.
13. Kamm W. and Goudert K., The streamlined car, *Motor Life Magazine*, New York, October 1954.
14. Kieselbach R. J. F., Streamlined cars in Germany, aerodynamics in the construction of passenger vehicles 1900–1945, *Kohlhammer Edition Auto und Verkehr*, Stuttgart, 1982.
15. Kieselbach R. J. F., Streamlined cars in Europe/USA, aerodynamics in the construction of passenger vehicles 1900–1945, *Kohlhammer Edition Auto und Verkehr*, Stuttgart, 1982.
16. Klemperer W. Luftwiderstandsuntersuchen an Automobilmodellen, *Zeitschrift für Flugtechnik und Motorluftschiffahrt*, Vol. 13, 1922, pp. 201–6.
17. Koenig-Fachsenfeld R., Vom Zeppelin auf Radern bis zur K-form, Introductory paper, *Proc. 4th Colloquium on Industrial Aerodynamics – Road Vehicle Aerodynamics*, June 1980.
18. Lay W. E., Is 50 miles per gallon possible with correct streamlining? *SAE Journal*, Vol. 32, 1933, pp. 341–54.
19. Ludvigsen K. E., The time tunnel – an historical survey of automotive aerodynamics, SAE paper 700 035, Detroit, 1970.
20. McCarthy M., Dam it – you'll spoil it, *Motor*, week ending March 6, 1976.
21. McDonald A. T., A historical survey of automotive aerodynamics, *Aerodynamics of Transportation, ASME-CSME Conference*, Niagara Falls, 18–20 June, 1979, pp. 61–9.
22. Morelli A., Aerodynamic basic bodies suitable for automobile applications, in: *Impact of Aerodynamics on Vehicle Design, Proc. International Association for Vehicle Design: Technological advances in vehicle design*; SP3, ed. Dorgham M. A., 1983, ISBN 0 907776 01 9.

23. Nouzawa T., Hiasa K., Nakamura T., Kawamoto K. and Sato, H., Unsteady-wake analysis of the aerodynamic drag on a notchback model with critical afterbody geometry, SAE SP-908, paper 920202, February 1992.
24. Scibor-Rylski A. J., *Road Vehicle Aerodynamics*, 2nd edn, Pentech Press, 1984.
25. White R. G. S., A rating method of assessing vehicle aerodynamic drag coefficients, MIRA rep. No. 1967/9. (see also: A method of estimating automobile drag coefficients, SAE 690189).

CHAPTER 5

COMMERCIAL VEHICLES

At one time, there seemed to be little point in trying to reduce the aerodynamic drag of commercial vehicles. Compared with private cars, most trucks, buses and delivery vans were relatively slow, with a high tyre rolling resistance, and their basic form, a large box on wheels, did not appear to be readily amenable to streamlining. Nowadays, however, this perception has changed; commercial vehicles on motorways or freeways regularly travel at the legal speed limit (or sometimes more!), which in parts of the world can mean speeds of up to 80 mph (130 km/h) for some categories including large buses. Furthermore, it has been realized that significant reductions in drag can be obtained by the simple expedient of rounding the front corners of the box. For a single-deck bus, for example, it is not particularly difficult to achieve drag coefficients in the region of 0.35: a value similar to that of a typical car. Figure 5.1

Fig. 5.1 An articulated truck with a number of drag-reducing features described later in the text; these include a cab-roof fairing, rounded mouldings on the trailer leading edges, side skirts, and turning vanes on the cab front corners.

shows a large articulated truck with a number of streamlining features which, as described later in this chapter, could be expected to reduce the drag by more than 30 per cent.

ECONOMIC FACTORS

The potential economic advantages of reducing the drag of large trucks has been shown to be considerable. In the late 1970s, Buckley *et al.*[1] estimated that fitting simple drag-reduction devices to all long-haul tractor-trailer trucks in the USA would result in a decreased fuel consumption of 1 billion gallons per year: 10 per cent of the fuel being used by such vehicles. Great care however has to be taken when assessing the potential for fuel savings. For example, if reducing the wind resistance merely allows the vehicle to travel faster, there may be little or no reduction in fuel consumption. This is likely to be the case for a large fully laden truck, where the speed may be limited by the drag and the power available rather than by legal limits. This aspect is discussed in more detail at the end of the chapter.

As described in Chapter 3, it is no simple matter to estimate the fuel-saving potential from wind-tunnel tests, since the effects of atmospheric wind turbulence and side-winds will significantly affect the drag. This is particularly true of articulated tractor–trailer vehicles, where the gap between the two units produces a high sensitivity to yaw angle. The influence of yaw angle on drag tends to increase with increasing vehicle length/width ratio, so, as may be seen from Fig. 2.16, most types of commercial vehicle tend to be more sensitive to side-winds than domestic cars.

Buckley's estimates of fuel savings were based on a very detailed analysis which took account of the effects of side winds, but since that time, streamlining and the fitting of aerodynamic devices to commercial vehicles has become widespread. It is, however, still possible to obtain significant further reductions in fuel consumption. Table 5.1 shows drag coefficient values for some basic types of vehicle of pre-1970 vintage compared with current typical values, and projected potential values obtainable by more radical or thorough approaches.

Table 5.1 Approximate ranges of drag coefficient values for various types of vehicle. Before the 1970s, little attempt at streamlining was made, and currently it is applied rather haphazardly. The predicted minimum values are quite possible, taking advantage of current knowledge, and do not require impractical or extreme measures.

Vehicle type	Pre-1970	Current	Probable near-future minimum
Medium-sized cars	0.4–0.55	0.28–0.4	0.25
Light vans	0.4–0.6	0.35–0.5	0.3
Buses	0.5–0.9	0.4–0.8	0.3
Large articulated trucks	0.7–0.95	0.55–0.8	0.4
Box truck and drawbar trailer	0.75–1.0	0.7–0.9	0.5

THE BASIC FORMS OF COMMERCIAL VEHICLES

Commercial vehicles come in a great variety of forms, but attention here will be focused on four types: the panel van shown in Figs. 5.2 and 5.3, the large articulated truck, the smaller non-articulated box truck, and the single-deck bus.

In the early days, most vehicles had a separate engine compartment ahead of the driving cab, but since the 1940s, forward control or cab-over-engine (c.o.e.) vehicles have become increasingly popular. In these, the engine is placed either behind or partially under the driver. One of the most notable smaller vehicles was the Volkswagen panel van (Fig. 5.2) which appeared in the late 1950s. In this vehicle, the rear-mounted air-cooled engine produced a relatively smooth-fronted brick shape, resulting in a drag coefficient of 0.45, which was better than for most cars of the period. Large European trucks also started to adopt a flat-fronted form with the engine below or behind the driver, but most intermediate-size vehicles tended to retain a vestige of a front engine compartment as illustrated in Fig. 5.3 (top left). This has been refined, firstly by adopting a double-slope front as in Fig. 5.3 (top right), and finally by producing the

Fig. 5.2 The original split-screen VW panel van of the late 1950s represented a major advance in small commercial vehicle design. The well-rounded front end and generally clean lines resulted in a C_D of 0.45 which was better than most cars of that period. The air-cooled rear-mounted engine eliminated the normal front radiator grille. This example is from 1965. (*Photograph courtesy of R. G. Dowson*)

Fig. 5.3 Panel van forms
With vestigial engine compartment (top left); double slope front (top right); simple raked front (bottom).

highly raked simple form as in Fig. 5.3 (bottom). Nowadays many panel vans and buses have a simple brick-like form with either a plane vertical, or raked front (Fig. 5.3 (bottom)). Large trucks do not yet conform to this configuration, although there has been some trend towards more streamlined and integrated shapes, as described later.

DRAG REDUCTION BY TREATMENT OF THE FRONT END

As was explained previously, the rounded front-end shapes of domestic cars result in very little if any drag being produced directly on the front surfaces. On a sharp-edged flat-fronted vehicle, however, the pressure over most of the front surface is positive, resulting in a large contribution to drag. Hoerner[7] quotes a pressure drag coefficient of 0.8 for *the front face* of a blunt circular cylinder. As shown in Fig. 5.4, however, the

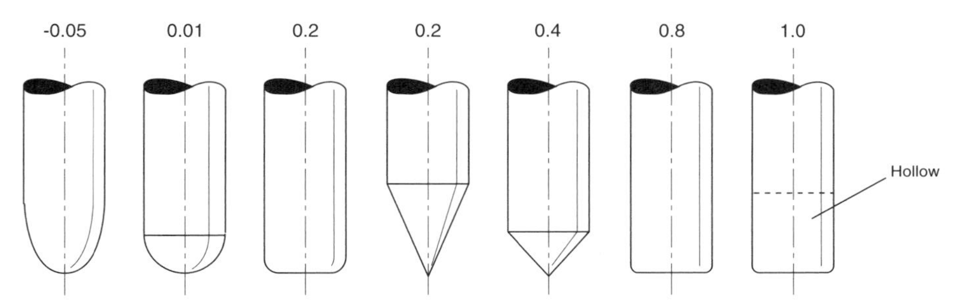

Fig. 5.4 Drag coefficients of the forebody pressure drag for a series of cylindrical bodies evaluated from measurements of the pressure distributions.
(*After Rouse and McNown*[12])

drag coefficient can be negative for a streamlined bullet shape, where the overall drag comes from the low pressure on the rearward-facing surface and from surface friction. Between the two extremes of the flat and bullet-shaped front face, the overall drag depends on the degree of rounding of the front edges. Figure 5.5 shows how quite dramatic reductions in drag can be obtained simply by rounding off these corners to reduce or remove the area of separated flow. This figure also illustrates how sensitive the drag of the rounded brick-like form is to Reynolds number, a factor that leads to considerable difficulties with wind-tunnel model testing.

It will be seen from Fig. 5.5 that the drag coefficient initially falls rapidly with increase in the radius of the front edge, but a point is reached where the drag curve tends to flatten out. This could be described as an optimum radius, since any further increase produces little benefit, at least in the zero yaw angle case. On the basis of published data, the optimum radius can be seen to depend on the length of the box; the shorter the box, the greater is the required radius. For a long bus shape, the optimum corner radius to face width ratio r/b is around 0.06 for full-scale cruising speed Reynolds numbers, whereas for a shorter block more typical of a panel van, it is closer to 0.1. This difference in the optimum radius is due to the fact that for the longer body, reattached flow occurs over a greater proportion of the length.

The idea of a simple optimum radius is unfortunately not appropriate for a real vehicle travelling along a road, because, as we have described earlier, the vehicle is generally subjected to varying intensities of cross-winds. It is therefore necessary to consider drag coefficient values for a wide range of yaw angles, not just the headwind case. It is found that as the yaw angle increases, a larger corner radius is required in order to inhibit separation. The so-called optimum or critical radius mentioned above is only relevant to the special case of a vehicle driving along with no cross-wind component. Nevertheless, since on average the side-wind component is not usually large, a corner radius to vehicle width ratio (r/b) of around 0.1–0.15 is probably adequate for practical purposes. This corresponds to something of the order of 30 cm (12 in) radius for a large truck.

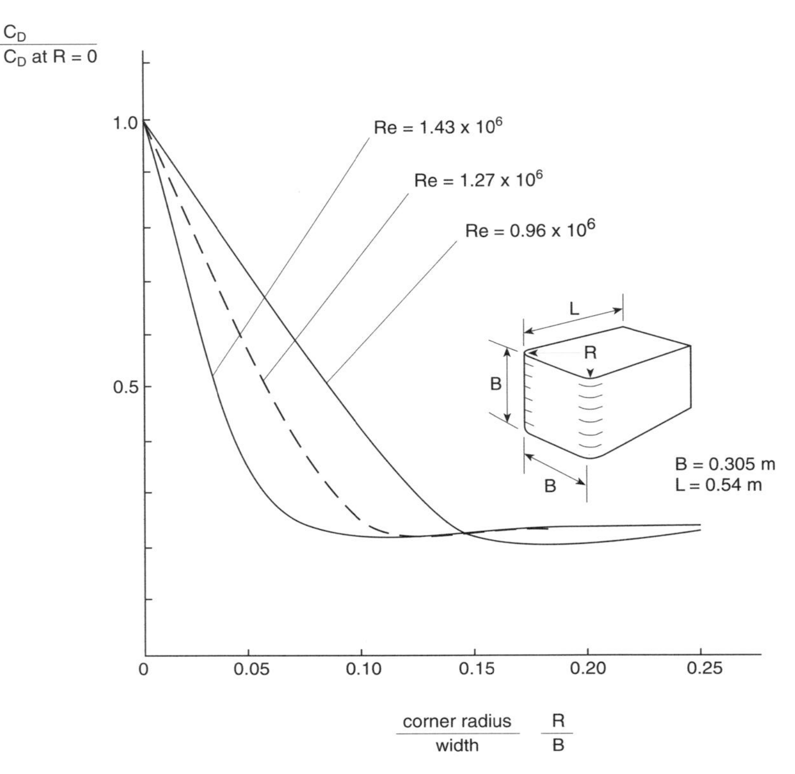

Fig. 5.5 The effect of front corner radius on the drag of a brick-shaped body at zero yaw angle.
(*Based on the data from Palowski*[10])

Although altering the front-end form of commercial vehicles by rounding the corners may seem a relatively simple design change, it required some modification to production methods, and perhaps more importantly, it required stylists and customers to come to terms with the new rounded and featureless shapes. The classic American engine-in-front 'big rig' beloved of trucking devotees is not compatible with low aerodynamic drag. Serious attempts at drag reduction also meant the abandonment of the cherished frontal ornamentation and styling features.

The smoothly rounded and ornament-free front end can be seen on most modern commercial vehicle designs, from the small delivery van (Fig. 5.3) to the large articulated tractor-trailer (Fig. 5.24). The old Volkswagen van shown in Fig. 5.2 is a good example; the well-rounded front end was almost devoid of features, although technological limitations dictated the use of flat rather than well-curved windscreen panels.

One geometric variable that might seem important is the rake angle of the front surface. Instinctively, it might be imagined that raking the front would make the vehicle more streamlined; however, Carr[2] compared the drag coefficients of various front-end geometries applied to a basic box shape, as illustrated in Fig. 5.6. As will be seen, the drag coefficient values are mostly very similar at around 0.21 to 0.25. Raking the front face back produces a small drag reduction, as may be seen, but inclining it forward produces a strongly adverse effect. Gilhaus[6] shows a drag coefficient reduction of 0.02 resulting from raking of the front of a bus back. More significant drag reduction can be achieved if extreme rake angles of around 60° to the vertical are used. There are some practical difficulties in using such large angles for commercial vehicles, although the Pontiac Trans-Sport shown in Fig 5.7 shows

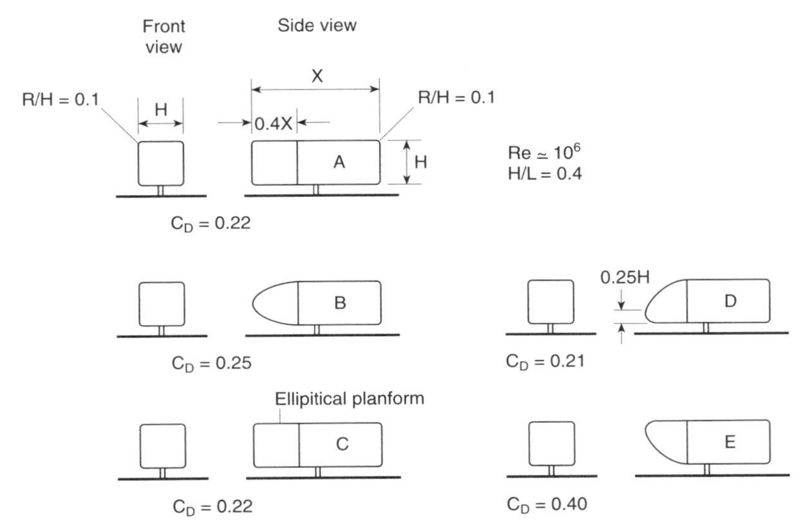

Fig. 5.6 The influence of front-end shape on the drag coefficient of rectangular bodies in ground proximity.
(*After Carr,*[2] *courtesy of MIRA*)

Fig. 5.7 A well-rounded front and a high degree of screen rake are apparent on this Pontiac Trans-Sport. The good overall shape combined with the flush glazing and continuous clean panel lines shows what is possible with a small van-like configuration.

that large rake angles can be used on a van-type shape. The drag coefficient is not very sensitive to the rearward rake angle, so the stylist has at least some scope for variations in the front-end form.

THE REAR END

Radiusing and tapering the rear end produces significant drag reductions. Hucho[8] quotes C_D reductions of 4–8 per cent for radiusing, and up to 20 per cent for side panel and roof tapering, with a combination of taper and radiusing producing up to 22 per cent reduction. Radiusing or tapering the rear end of commercial vehicles, however, is not always entirely practical. In many cases, the load container needs to be in the form of a rectangular box; furthermore, a tapered rear end hampers loading access. Nevertheless, as discussed in more detail below, there are some special cases such as buses where a significant degree of rear-end rounding or tapering is possible. The little Renault Clio van shown in Fig. 5.8 is a good example of rear-end tapering and radiusing applied to a small commercial vehicle.

The following sections deal with means of drag reduction on particular types of commercial vehicle.

Fig. 5.8 The Renault Clio van
Showing that a well-tapered and rounded rear end is a practical proposition.

PANEL VANS

Most panel vans now have one of the general forms shown in Fig. 5.3, but the trend is towards the the fully sculpted form with a smooth nearly featureless appearance, as in Figs 5.3 (top right), 5.3 (bottom) and 5.7. The radiator intake is often almost hidden at the bottom. Producing really smooth rounded shapes economically has required a significant change in production techniques for most manufacturers. The basic form has gradually been refined towards ever smoother contours, using features such as flush glazing. In respect of progress in drag reduction, there seems little scope now for improvements over the best contemporary examples in terms of basic upper-body shape. Any further drag reduction will require attention to details such as wing-mirrors and wheel trim, and improvements to the underbody flow, the latter being the most important and potentially fruitful area to work on. C_D values of around 0.3 are a reasonable target for vehicles of this type.

At the time of writing, the trend for panel vans appears to be towards the simple raked front-end single-box form illustrated in Fig. 5.3 (bottom), although the more conservative alternative forms illustrated in Figs 5.3 (top left) and 5.3 (top right) are also still popular. In both of these latter cases, the use of a step in the raking of the front face would be expected to generate a certain amount of trailing vorticity, resulting in a drag penalty. Figure 5.9 shows results from tests on a simple wind-tunnel model

Fig. 5.9 The effect of front-end shape
The model with front screen and engine cover raked at different angles has a C_D of 0.823 compared to 0.814 for for the model with a singly-raked face. By rounding the front and rear corners it was possible to reduce the coefficient to 0.421 for the second case. The high C_D values are due to the sharp leading corners.
(From tests conducted at the University of Hertfordshire by J. Comer)

with sharp leading corners. This figure indicates the order of the drag penalties relative to the simple raked shape. The high drag coefficient values obtained in the experiment were due to the sharp corners; with rounded front and rear corners, the drag coefficient dropped to 0.421.

Most panel vans are rear-loaded, albeit sometimes with an additional option for side-loading. Rear-loading precludes the use of significant rear-end taper, except on the underside, but a reasonable amount of rounding is possible.

BUSES

For buses, the brick-like form has been almost universally adopted since the 1970s. In many cases, the engine is mounted at the rear to lower the driving position. Before the importance of aerodynamics was appreciated, many European designers tended to favour completely blunt front faces, with little radiusing of the corners. This produced extremely high drag coefficients of around 1 or more. As with the panel van, significant reductions in drag can be obtained simply by rounding the front edges; Fig. 5.10 shows a typical more modern design.

Figure 5.11 shows the progressive effects of modification of the front-end shape. As with the panel van, raking the front screen to modest angles produces no dramatic effect, but does give some scope for purely stylistic design variations. Where a roof-level ram-air intake is required for cooling and ventilation purposes, the rounded edge form shown in Fig. 5.12 will give a relatively small drag penalty.

The drag of large buses of the rounded brick form will generally be higher than for a panel van because of the greater length to width ratio of a typical bus, which produces a greater surface friction drag. One advantage of buses however is that they are normally

Fig. 5.10 Bus with a well-rounded front face
This design contrasts with the angular box-like forms of the previous generation.

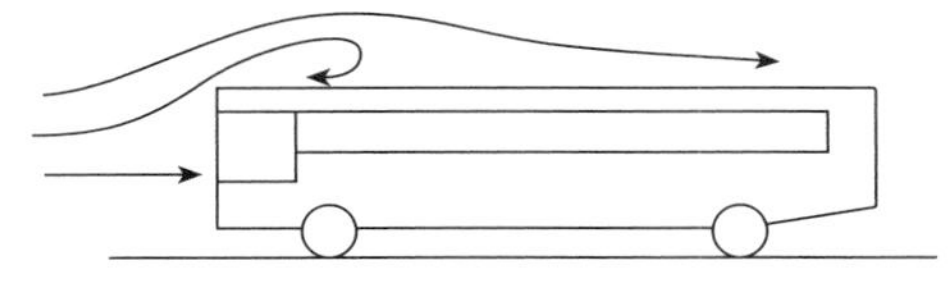

Sharp leading corners, $C_D = 0.88$

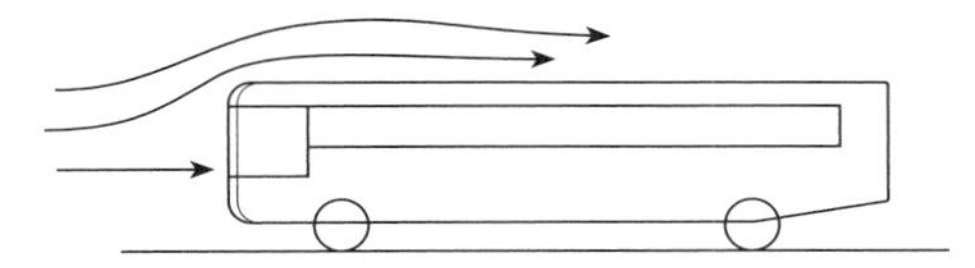

Rounded leading corners, $C_D = 0.36$

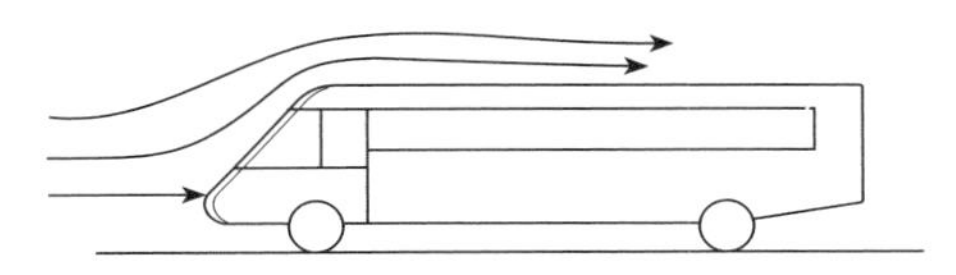

Rounded and raked leading face, $C_D = 0.34$

Fig. 5.11 The influence of front-end rounding and rake on the drag coefficient of a bus. (*After Gilhaus*[6])

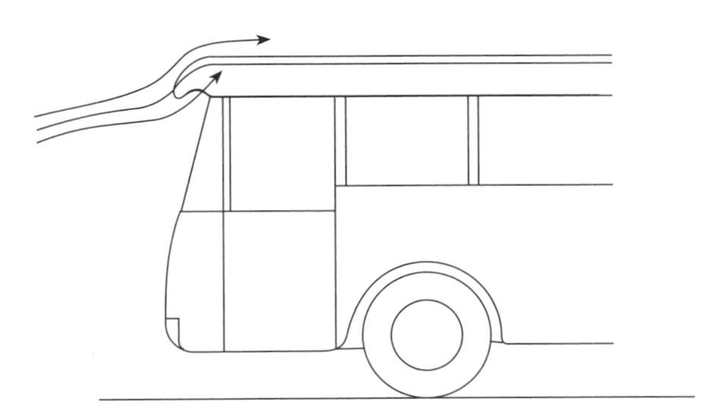

Fig. 5.12 The roofline is sometimes extended forwards on buses to provide a ram air intake for ventilation or air conditioning. Suitable rounding of this extension minimizes the drag penalty.

side- rather than rear-loaded, which means that some degree of rear-end taper and radiusing can be tolerated. In the 1930s a number of streamlined buses were designed based on the ideas of Jaray and Kamm; two examples are sketched in Fig. 5.13. These vehicles show considerable rear-end taper and rounding both in plan and side views.

The modern angular brick-shaped bus offers significant advantages in terms of manufacturing costs, but much of the resistance to more streamlined forms no doubt

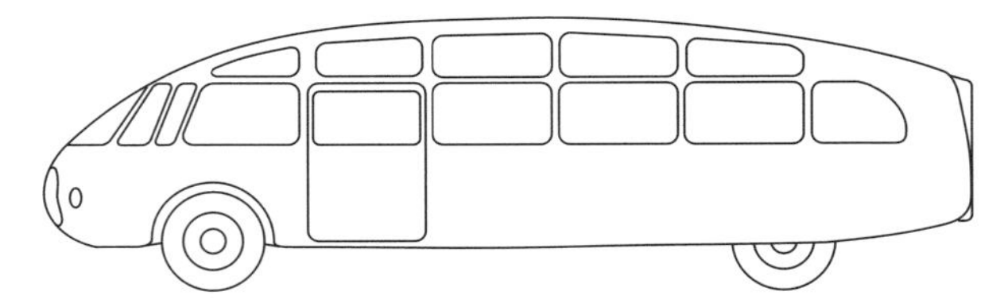

Fig. 5.13 Some designs for streamlined buses from the 1930s. The large degree of rear taper looks odd by contemporary standards, but is not impractical.
(*After Koenig-Fachsenfeld*[9])

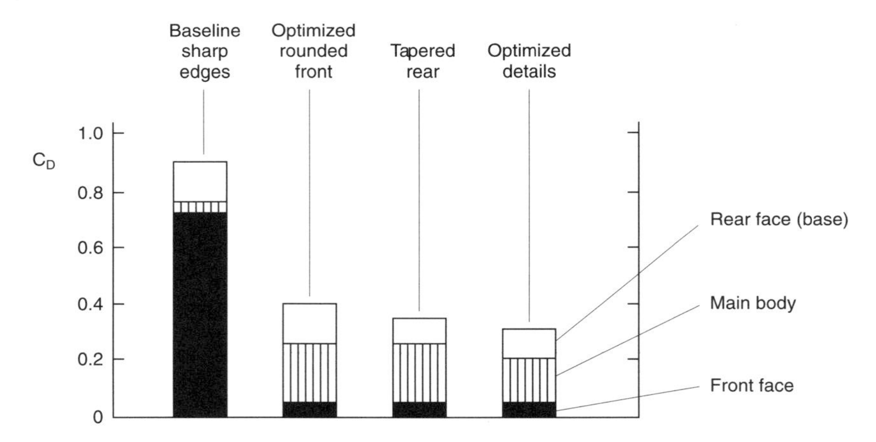

Fig. 5.14 Drag reduction methods applied to a bus starting with a sharp-edged brick shape.
(*Based on data in Gilhaus*[6])

comes from styling perceptions. A small amount of rear-end taper would probably be acceptable, particularly on the underside. Wind-tunnel studies by Gilhaus[6] indicate that a drag coefficient of 0.3 may be obtainable with an optimized bus shape. Figure 5.14, which is based on data from Gilhaus, shows the effects of various aerodynamic design improvements, starting from the baseline of a sharp-edged brick shape. The final configuration gives a C_D of 0.3, which represents a realistic target for a low-drag bus. Note that the very significant improvement over older designs is obtainable without a major change in styling or general configuration. This low-drag bus would have well-rounded front and rear corners, with the front face raked. The roof and underside panels would be inclined at the rear to produce a degree of taper.

HEAVY TRUCKS

In economic terms, articulated trucks or tractor-trailer units are the most important category of commercial vehicle. They are also the most difficult to deal with in respect to aerodynamic design optimization, because their geometry can vary in a number of ways. The ability of the tractor unit to turn relative to the trailer means that there is normally a large gap between the upper parts of the two units. In addition to this inherently variable geometry, the tractor may pick up a variety of shapes of trailer, or may deposit a container unit and return with an open flat-bed trailer. Because of both the economic and technical interest, the various aerodynamic characteristics of these vehicles are dealt with in some depth here. It should be noted that most of the drag reduction techniques mentioned below are equally applicable to rigid-box trucks.

INTERACTION BETWEEN TRACTOR AND TRAILER

The most common form of large articulated truck has the simple two-box arrangement with tractor/cab unit and a closed-box trailer section, as illustrated in Fig. 5.15. Unless corrective action is taken, the stepped shape in side elevation (and

Fig. 5.15 A typical older-style articulated truck with no aerodynamic devices apart from small turning vanes on the front corners in line with the radiator grille.

sometimes in plan view as well) tends to create a relatively high drag. As shown in Fig. 5.16(a) the flow will separate over the top of the driving cab, impinge on the front of the carrying box, and then separate again along its front edges. This behaviour of the flow explains why there is a strong interaction between the shape of the tractor and trailer. The shape of the tractor will influence the way in which the flow impinges on the trailer. Rounding the front edges of both boxes will normally produce a reduction in drag, but in some circumstances rounding the cab unit can actually increase the drag. The rounded cab form in Fig. 5.16(a) causes the flow from the cab top to impinge on the trailer front face producing a high-pressure stagnation region there, followed by a strong separation at the upper edge of the trailer box. In contrast, the sharp-edged cab roof in 5.16(b) deflects the flow towards a relatively

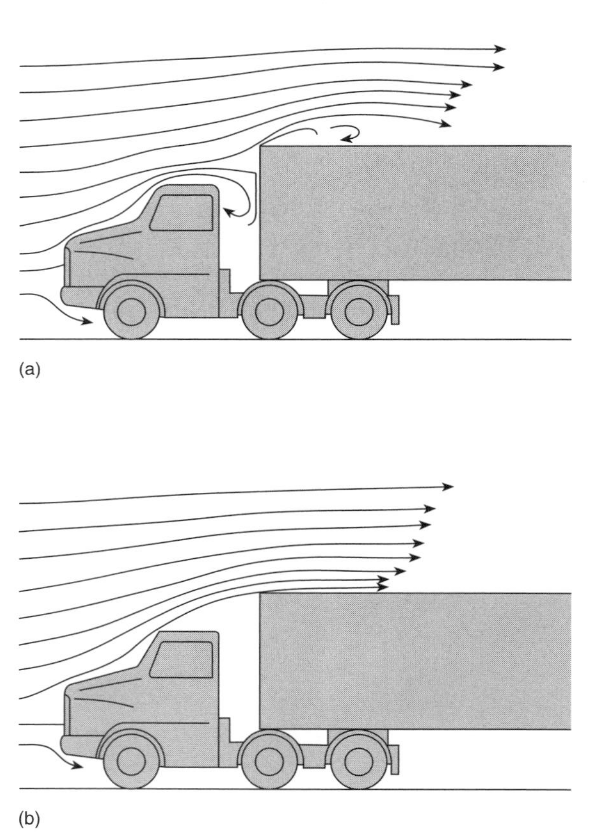

Fig. 5.16 The interaction between tractor and trailer flows
The rounded tractor front form in the upper sketch causes the flow to impinge on the trailer front face resulting in a greater drag than for the arrangement with the sharp-cornered cab shown in the lower sketch. The sharp edge causes the flow to separate early, and the flow is thus directed over the front face of the trailer.
(*Based on flow visualization photographs in Gilhaus*[5])

smooth reattachment on the top surface of the trailer. The overall drag coefficient for the rounded cab model was thus 0.73 as compared with only 0.6 for the sharp-edged model. This example illustrates the dangers of generalization. Rounded edges normally reduce drag, but there can be exceptions. These complex types of interaction make the articulated vehicle a very difficult case to study.

REYNOLDS NUMBER PROBLEMS

One problem with trying to optimize the shape of large vehicles by means of experimental tests is that they require extremely large wind-tunnels. A tunnel that will take a 1/4 scale car model will only accommodate a 1/8 or 1/10 scale truck. At this scale the maximum Reynolds number obtainable is low enough to be in the region where data are still very sensitive to Reynolds number. Wind-tunnels designed for full-scale car work are often too small to take a full-size large truck without incurring problems due to high blockage. It is quite common therefore to test large-scale (typically 1/2 scale) models in full-scale car tunnels.

OPTIMIZING THE SHAPE OF LARGE ARTICULATED VEHICLES

Because of the economic importance of large trucks, there have been many experimental studies of their aerodynamic characteristics. Unfortunately because so many variables are involved, the result has been the creation of a vast and somewhat confusing array of data. Rather than get bogged down with the details of current designs which will soon be obsolete, it will be more helpful to look at what is involved in the design of a low-drag vehicle. The reasons for deficiencies in current or older designs will then become apparent.

Figure 5.17 shows a proposed optimized design for a vehicle constrained by the requirement to carry a sharp-edged rectangular box container. The best that can be

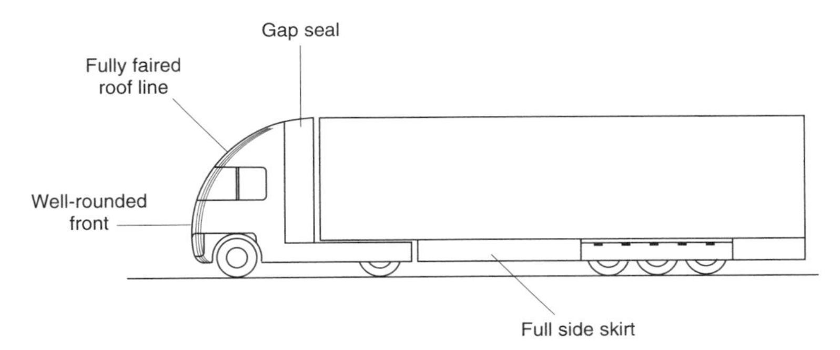

Fig. 5.17 Design features of a well-streamlined articulated truck
Drag coefficient values of around 0.4 to 0.5 should be possible with careful attention to detail design.

done is to turn the tractor unit into a streamlined front-end fairing. The basic shape here corresponds roughly to a cylinder with a streamlined nose for which Hoerner's C_D value is 0.2. Of this, about 0.15 is due to the low pressure on the rear (base) surface, and about 0.05 to surface friction. For a practical vehicle, the actual drag coefficient will be considerably greater than this. The front surface will not be a perfect bullet shape, the side, top and under surfaces will not be absolutely smooth, and the wheels, cooling flow etc. will make major drag contributions; nevertheless, the shape in Fig. 5.17 is basically the same as for the rounded-front bus, for which a drag coefficient of 0.35 is readily realizable. For a large truck the figure would be larger because in most cases it would not be possible to achieve smooth side surfaces. Also, if we are truly constrained to use a sharp-edged box form for the load container, then there is no possibility of reducing drag by rounding or tapering the rear.

THE INFLUENCE OF THE TRACTOR-TRAILER GAP

The first practical obstacle to low drag on an articulated vehicle is the clearance gap that has to be provided between tractor and trailer. If the gap were very small, then it would not be expected to have much effect; both the rearward-facing surface of the tractor and the forward-facing surface of the trailer would see roughly the same pressure, and the forces produced would cancel. At the other extreme, with a very large gap, the configuration tends to behave as two separate bodies, with roughly twice the drag of a single body. From this, it would be expected that the drag of an articulated vehicle would rise with increasing gap size. To interpolate between extremes in this way is rather dangerous, because in the region of real practical interest, the flow structure is quite complex; nevertheless, published data do show that drag rises significantly with gap width. For one cab-over-engine vehicle, Buckley *et al.*[1] show an increase in drag coefficient at zero yaw angle of 0.15 for a gap increasing from 61 cm to 183 cm.

The tractor-trailer gap has an even greater importance when the vehicle is yawed relative to the wind direction. As may be seen from Fig. 5.1, articulated vehicles are much more sensitive to wind direction than other types, due to the influence of cross-flows through the gap. Figure 5.18, which is based on flow visualization studies, shows the effect of the gap flow in a cross-wind. The gap between tractor and trailer is often quite unnecessarily large, particularly on American vehicles (Fig. 5.19).

Having identified the gap as a major contributor to drag, it is possible to devise designs that would eliminate or reduce the effective gap. Various forms of flexible or rubbing seal have been proposed, but the simplest method is to extend the side panels of the tractor unit backwards as shown in Fig. 5.20. This requires that the tractor cab should be as wide as the trailer unit, or that the side extensions should flare out. On current designs the cab is often narrower than the trailer, but there has been a general trend towards wider cabs. With rigid side extensions, and the pivot distance from the front face equal to one half of the trailer face width, a 90° articulation angle is permissible without the trailer fouling on the extensions. With

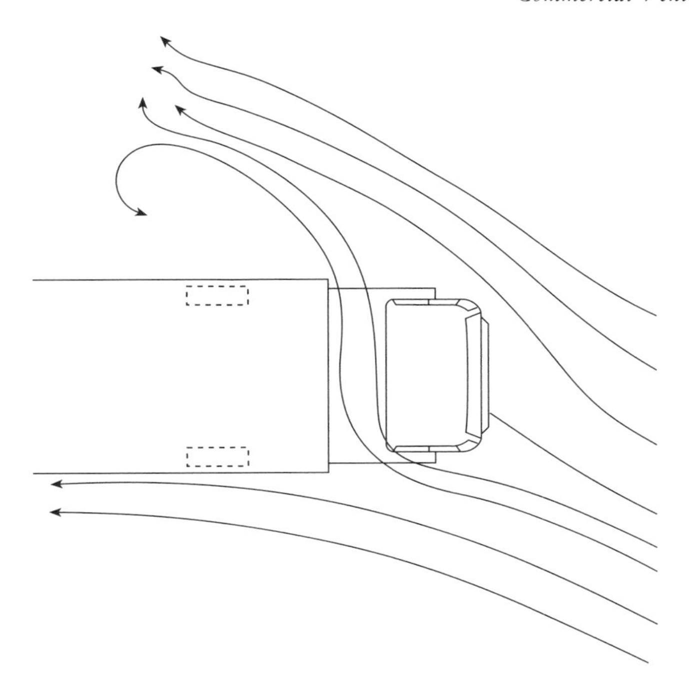

Fig. 5.18 The flow through the tractor-trailer gap in a side-wind, based on flow visualization studies in Hucho[8].

Fig. 5.19 The front-engined truck configuration still popular in the USA, where it is known as the conventional type. Note the large gap and the extreme length of the trailer, which would produce significant surface friction drag.

the pivot position further forward, or if greater articulation angles are necessary, the extensions would need to be hinged or flexible. Gap fairings of this type have been shown to produce drag coefficient reductions of around 0.05 at zero yaw, with improvements of up to 0.23 at 10° yaw. The actual degree of reduction will of course be affected by a number of geometric features, so no general figure can be quoted.

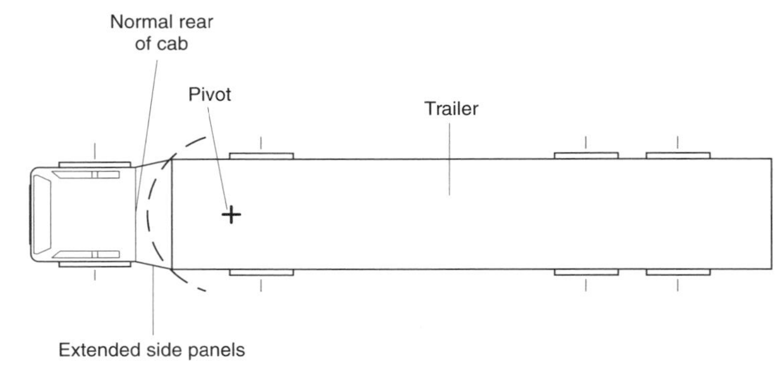

Fig. 5.20 Gap sealing by extension of the side panels of the cab rearwards (top view).

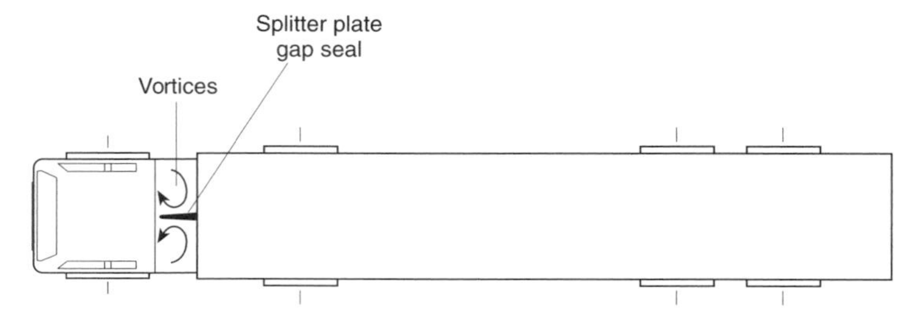

Fig. 5.21 Centre-line gap seal with standing vortices.

As an alternative to a complete gap fairing, Buckley *et al.*[1] looked at the effect of a vertical plate or 'gap seal' on the centre-line (Fig. 5.21). The plate prevents cross-flows through the gap at high yaw angles, and stabilizes the flow by generating a pair of relatively stable vortices. Although a little less effective than the gap fairings described above, the order of drag reduction is found to be similar, being again in the region of 0.05 at zero yaw, and increasing in effectiveness with yaw angle. Gap seals require a certain amount of design ingenuity in order to accommodate the angular movement of the trailer. Various patent designs of partial gap seal are also available. They are known as vortex stabilizers, because they generate a pair of standing vortices, similar to those generated by the single gap seal, as illustrated in Fig. 5.21.

Another form of vehicle that suffers from the gap problem is the combination of box truck and trailer unit. Sealing the gap here represents a much greater problem, and there is no obvious simple and truly practical solution.

UNDERSTANDING AND IMPROVING EXISTING DESIGNS

Having considered an 'ideal' configuration, it is now easy to see the defects in older designs of articulated truck. There are two basic configurations, the cab-over-engine type shown in Fig. 5.15, which has been the standard form in Europe for many years, and the 'conventional' or front-engined arrangement shown in Fig. 5.19, which is still popular in the United States. Vehicles designed before the importance of aerodynamics was recognized often had very high drag coefficients, with values ranging between 0.7 and 0.9 in the zero yaw case. Rounding and generally streamlining the front end of the tractor unit can bring this down, but as is evident in Fig. 5.16, rounding the front of the cab can sometimes actually make matters worse by increasing the exposure of the trailer. In both forms, the front edges of the trailer box may be exposed to the flow, in which case, these will also need to be rounded, which may not always be a practical proposition. For low drag it is necessary to shape the front surface so that the flow just meets the edges of the trailer, and this can only be done on the basis of largely trial and error wind-tunnel tests. Purely intuitive attempts at streamlining the tractor unit are likely to produce disappointing results.

DRAG-REDUCING DEVICES: 1 - CAB-ROOF DEFLECTORS AND FAIRINGS

Considerable reductions in drag can be effected if the separated flow from the cab roof is directed to attach on or near the leading corners of the rear box. The resulting flow field in profile then approximates to that of the ideal form in Fig. 5.17.

The simplest method illustrated in Fig. 5.22 (left), consists of a deflector plate mounted on the cab roof. An alternative is a cab-roof fairing as shown in Fig. 5.22

Fig. 5.22 Cab-roof devices
A deflector plate (left) is a cheap simple device which can be quickly fitted, and in some cases can be folded flat when running without the trailer, or with low loads. A cab-roof fairing (right) which tapers towards the front in plan view to encourage flow attachment in strong cross-wind conditions. The relatively sharp edge radius helps fix the separation line in cross-winds, and thus aids stability. This unit has provision for height adjustment.

(right). The fairing is generally more effective than a deflector because it can direct the flow accurately both in side and plan views. It also reduces the extent of recirculating flow in the gap region. In the case of the simpler non-articulated truck, the cab-roof fairing can extend right up to the rear box, and in this case, the overall shape then approximates to the ideal rounded-front single-box form.

One potential advantage of the deflector is that it is possible to design it so that it can be lowered when the tractor unit is run without its trailer, or when a container is off-loaded, leaving the trailer in open flat-bed form for the return journey. The relatively poor performance of deflectors in cross-winds is, however, a significant disadvantage.

There are many forms of cab-roof deflector and fairing, some of which seem to have been designed on the basis of faulty intuition rather than an understanding of aerodynamics. Any device featuring concave surfaces probably falls into this category. The false idea that you can scoop air up persists in popular imagination. In practice it is invariably better to use a convex form.

The most popular form of cab-roof fairing is shown in Fig. 5.22 (right). The plan-view shape narrows towards the front, and should produce attached flow over a wide range of relative wind flow directions. The fairly tight radius along the top edges is not ideal for drag minimization, but fixes the separation position, helping cross-wind stability. An alternative form, which blends smoothly into the normal shape of the cab, is shown in Fig. 5.23.

Figure 5.24 shows how the deflector produces a negative (relative to atmospheric) pressure over the upper portion of the trailer front face, in contrast to the positive (drag-producing) pressure in the case without a deflector. Drag reductions of up to 30 per cent have been measured using such fairings (see Hucho[8] and Garry[4]), but figures such as this need to be treated with caution, because the magnitude of the reduction depends on how bad the drag was in the first place.

Fig. 5.23 A full-width fairing blended to the contours of the cab roof.

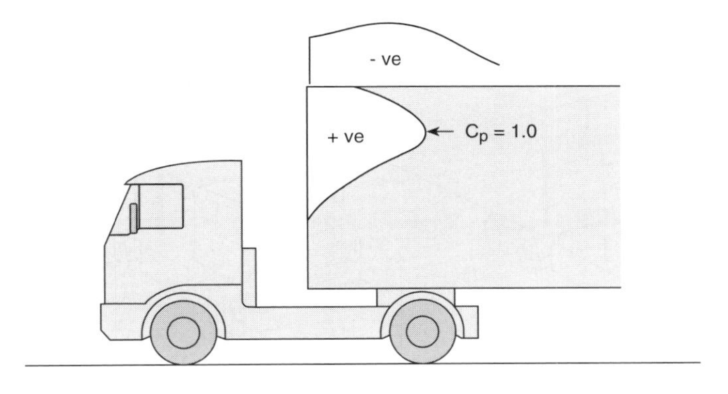

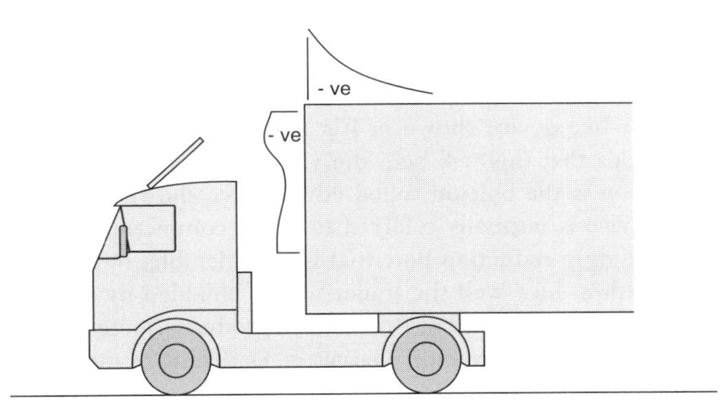

Fig. 5.24 Pressure distributions on the front of a trailer unit with and without a cab-roof deflector
Note how the pressure on the front face of the trailer changes from positive to negative with the fitting of a deflector. This means that the front of the trailer will make a negative contribution to drag.

With a cab-over-engine configuration, and well-designed roof and gap fairings, the vehicle will approximate to the ideal arrangement of Fig. 5.17. Buckley's data show that a combination of the roof fairing and a gap seal can produce a drag reduction of up to 35 per cent starting from a traditional non-streamlined vehicle. Really effective drag reduction can only be assured by wind-tunnel testing.

DRAG-REDUCING DEVICES: 2 – TRAILER-MOUNTED UNITS

As described previously, on many older designs, a great deal of drag was produced by flow separation from the leading corners of the rear box. Various

Fig. 5.25 Two trailer-mounted bolt-on drag-reducing devices: the splitter panel (left); the 'windcheater' (right).

proprietary 'fixes' for this problem have been marketed in the form of bolt-on panels which can be attached to the front face of the rear box. Some of these, such as the splitter-like device shown in Fig 5.25 (left), seem to have been based on the fallacious idea that this will help the vehicle cleave through the air. A much better solution is the bolt-on rolled-edge device shown in Fig. 5.25 (right). This ingenious device is normally referred to by its commercial name of 'windcheater'. The drag reduction potential is considerable, but the actual amount will depend on how well the trailer unit is shielded by the tractor. One advantage of the windcheater is that it normally produces some drag reduction regardless of the shape of the tractor; Garry[4] gives details of experimental investigations of such devices.

The magnitude of the drag reduction that can be obtained with these popular add-on devices depends greatly on the original geometry of the unmodified vehicle. Researchers report a variety of different values, although the cab-roof fairing and windcheater devices are usually seen to be the best choice, particularly in cross-wind conditions.

BOLT-ON TURNING VANES

One 'fix' that is sometimes applied to existing designs is a turning vane attached to the front corners. These vanes have a similar effect to radiusing the corners, and are useful when the basic vehicle has small radius corners. Hoerner[7] reports a drag coefficient reduction from 0.71 down to 0.26 for turning vanes applied to the front face of a blunt-nosed body of rotation, and a reduction from 0.58 to 0.27 when applied to a blunt rear face, as illustrated in Fig. 5.26. Such vanes are sometimes used on the leading corners of large articulated trucks; vanes may just be seen in Fig. 5.15. On the cab front, their use is usually restricted to the region

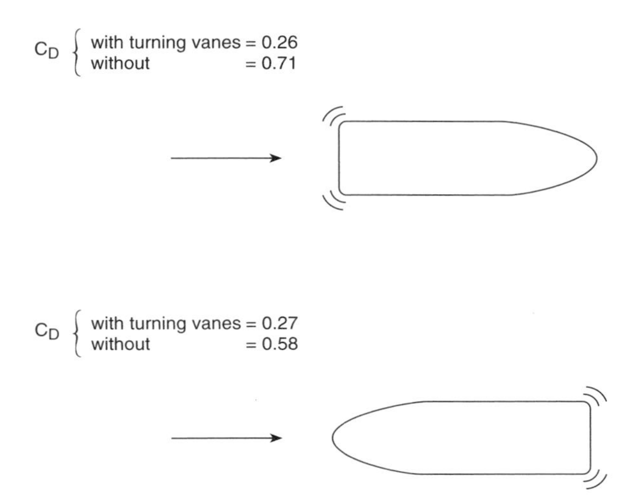

Fig. 5.26 The effect of turning vanes on the drag of a two-dimensional body at Re = 6×10^5.
(*After Hoerner*[7])

below the windscreen, because of their influence on visibility, although small vanes are sometimes fitted on the A-posts. In common with radiusing, the effectiveness of vanes depends on how the tractor unit wake interacts with the trailer box. Buckley *et al.*[1] reported a C_D reduction of 0.05 at zero yaw angle for such an arrangement applied to a sharp-edged cab. The reduction increased to 0.25 at 15° yaw.

Turning vanes can also be used as a 'fix' on the leading corners of sharp-edged trailer units, although Garry's work[4] suggests that the solid rounded front edges of the 'windcheater' device are more effective. Despite the experimental results quoted in Fig. 5.26, turning vanes have not shown much if any advantage when applied to the rear corners of road vehicles.

Add-on devices such as those described in the preceding paragraphs are currently in common use, but in the long term, one might expect that truck manufacturers would eventually produce more aerodynamically efficient designs. Although there is a clear trend in this direction, there is still the problem that a tractor unit can have a variety of trailers fitted. It is thus impossible to design a universally effective aerodynamically optimized tractor. The situation is not helped by the fact that trailers and tractor units are usually produced by different specialized companies. Manufacturers can provide ideal cab shapes for a range of standard trailer shapes, but add-on devices will continue to be required for use with non-standard trailer configurations, or when the operator may wish to use a variety of trailers behind the same tractor. Most cab-roof devices now incorporate a degree of geometrical adjustment in their design.

SIDE PANELS

Drag reductions may also be obtained by use of side fairings or skirts, which are illustrated in Fig. 5.27. Wind-tunnel studies at the University of Hertfordshire using the 1/15 scale model shown in Fig. 5.27 gave a measured drag reduction of 6.45 per cent (from a baseline C_D of 0.71) in the wind-on (zero yaw) case for the combination of tractor skirts and side guard panels illustrated. In this study, various combinations of side panel and skirts were tested, including full-length panels including rear wheel covers. The results, shown in Table 5.2, are in line with other

Table 5.2 Percentage change in drag coefficient from various arrangements of side panels on the trailer unit of an articulated truck (author's data).

Side panel arrangement	Percentage change in C_D from baseline vehicle value	
	Wind-on	15° yaw
Tractor unit skirts	5.4	8.4
Rear wheel fairings	1.1	3.8
Side guard fairings and tractor skirts	6.45	8.8
Full-length side fairing including rear wheel covers	7.0	9.3

Fig. 5.27 A wind-tunnel model fitted with tractor-unit skirts and partial side panels. Note the rectangle on the front face formed by strips of sandpaper, intended to provoke transition of the boundary layer to turbulent flow (see Chapter 11).

published data. It will be seen that the effectiveness of side panels increases with yaw angle, which is no doubt due to the fact that they inhibit cross-flows. Due to the small scale of the model, and correspondingly low Reynolds number, the results shown may not relate well to the full-size vehicle, although it is expected that the trends would be reproduced.

In many countries, there is now a requirement that guard-rails should be fitted under the trailers of articulated vehicles to prevent smaller vehicles being crushed underneath, and these rails provide a convenient attachment for the side panels. The regulations, however, specify minimum requirements for the strength and rigidity of such panels.

OTHER METHODS

Air dams similar to those used on cars or downward extensions of the front bumper (fender) are commonly found on more recent designs. Experimental investigations into the effectiveness show mixed results. To produce a reduction in drag, they need to be combined with skirts, at least for the tractor unit, otherwise, with the large ground clearance, the air dam will simply represent an increase in frontal area and a source of separated flow.

There may be some scope for improving the underbody flow of the tractor unit, but the cost of adding a panel under the trailer unit would almost certainly be prohibitive in most applications.

Turning vanes and extensions on the trailer rear end have been tried, and although some drag reduction can be obtained, most of such solutions are generally considered impractical. Dare-Bryan[3] describes a tapered roof panel and tapered rear side skirt valances fitted to commercial fleet rigid trucks as part of a drag-reduction package.

Experiments have been conducted (Buckley *et al.*[1]) to determine the effectiveness of active boundary layer suction or blowing to aid attached flow. Suction is used to remove the low-energy boundary layer flow, and thus encourage attachment. Blowing a thin jet of high-energy air into the boundary layer achieves the same effect. Significant drag reductions were found to be possible by these methods, but the mechanical complexity makes them unattractive, and the energy consumption of the pump has to be set against the fuel savings achieved. The demand for high aerodynamic efficiency would have to be very much greater before such extreme methods would attract serious interest. In the long term, the likely introduction of an energy tax and dwindling fuel reserves might eventually make it necessary to consider such techniques.

DISTRIBUTION OF DRAG BETWEEN TRACTOR AND TRAILER

Buckley *et al.*[1] measured the separate contributions to drag from the tractor and trailer units. For a 'conventional' or front-engined tractor coupled to a 3.66 m (12 ft) high trailer, the tractor unit produced a drag coefficient of 0.43, and the trailer 0.32. By contrast, for a cab-over-engine tractor with 30 cm (12 in) radius front edges coupled to a similar trailer, the tractor unit was found to contribute 0.51 to the drag coefficient, with only 0.11 coming from the trailer unit. The latter figure is less than the expected total of drag contributions from the rear face of the trailer and trailer surface friction drag. Clearly, therefore, the front face of the trailer in this case is subjected to a low pressure which is producing a forward thrust (negative drag) component. This is consistent with the pressure distribution shown in the second picture of Fig. 5.24, which is based on experimental measurements, and corresponds quite closely to the ideal configuration of Fig. 5.17, where the tractor effectively shields the trailer unit.

DRAG BREAKDOWN AND POTENTIAL FOR FURTHER REDUCTIONS

Table 5.3 gives a rough estimated breakdown of the expected drag coefficient contributions at zero yaw angle for a good modern design of articulated truck with a cab-roof fairing. Buckley *et al.*[1] found that the base pressure drag coefficient was around 0.1 for almost any front-end configuration. From theoretical considerations it can be estimated that for a 15 m long vehicle with absolutely smooth continuous surfaces, the surface friction drag coefficient at 100 km/h (62 mph) would be approximately 0.05. For realistic surface roughnesses, and taking account of the very rough underbody flow, this should be increased to around 0.15. The wheels might be expected to contribute another 0.1, with a further 0.1 coming from various sources such as the cooling system, and the gap. The degree of rounding used on current vehicles when coupled with a roof deflector or fairing should give a contribution of around 0.1 to 0.2 for the pressure drag on the front surfaces. This gives a total of 0.55 to 0.65, which is fairly typical for a well-faired and rounded cab-over-engine vehicle.

It would probably not be practical to go for the kind of bullet-shape front required for zero front-face pressure drag, but with a significant increase in rounding and smoothing it should be possible to reduce the front pressure drag coefficient to less than 0.1. With careful design, particularly of the underbody, it should also be feasible to reduce the surface friction drag contribution to around

Table 5.3 Contributions to the drag coefficient of an articulated truck at zero yaw angle.

Contribution	C_D
Base pressure	0.1
Pressure drag on front	0.1–0.2
Surface friction drag	0.15
Wheels	0.1
Cooling system, gap and other minor items	0.1
Total	**0.55–0.65**

0.1, and the minor items to 0.05, making a C_D value of 0.4 a reasonable target, with 0.3 a realistic possibility. Indeed Dare-Bryan[3] reports a measured C_D of 0.33 for a rigid truck fitted with a full retrofit package of drag-reduction panels. The fact that tractor and trailer units are normally designed and manufactured independently by different manufacturers currently inhibits good aerodynamic design, which requires an integrated approach.

FUEL SAVINGS

A general discussion of the relationship between drag coefficient reduction and fuel saving has been given in Chapter 3, but it may be helpful to consider the specific application to large trucks in more detail here. In one respect, estimating the fuel savings for a large truck is easier than for a domestic car, as trucks spend a relatively high proportion of their time cruising at their legal maximum speed, which in most countries is around 90–100 km/h (56–62 mph). At this speed, with drag coefficients typically in the region of 0.65, the aerodynamic drag represents approximately half of the total resistance, the other half being due to tyre rolling resistance; the proportion depends on how heavily the vehicle is loaded, since this affects the rolling resistance.

An interesting and extremely useful set of fuel consumption and drag measurement tests were conducted by Buckley *et al.*[1] Three articulated trucks were driven through the southern states of America from Maryland to San Diego California and back, a total distance of 9332 km (5800 mi), at an average speed of 91.3 km/h (56.7 mph). The vehicles were nominally identical 1974 model International Harvester C04070 cab-over-engine tractors, with 13.7 m (42 ft) long by 4 m (13.25 ft) high trailers. One was assigned as an unmodified baseline model; another was modified by the addition of a proprietary cab-roof deflector plate and a 'vortex stabilizer' in the gap. The third unit was fitted with a cab-roof fairing and gap-sealing arrangement developed by the University of Maryland. A 1/8 scale wind-tunnel model was also tested in the three configurations. Table 5.4 is based

Table 5.4 Comparison of drag reduction and fuel savings for two sets of devices.

Device	C_D (wind-tunnel, zero yaw)	Percentage drag reduction from baseline	Percentage drag reduction, wind-averaged (estimated)	Percentage fuel saving from baseline (measured)
Baseline vehicle	0.74	—	—	—
With deflector and vortex stabiliser	0.56	24.3	14.2	6.2
With fairing and gap seal	0.51	31.1	26.0	11.8

(Based on data in Buckley *et al.*[1])

on their results. Included in the table are values of the wind-averaged drag coefficient reductions, which are based on wind-tunnel data corrected for the effects of an average 11 km/h (7 mph) wind blowing from any direction, as outlined in Chapter 3. The wind-averaged drag reductions are smaller than for the basic zero yaw angle wind-tunnel data, because of the reduced effectiveness of the devices when the vehicle is in a cross-wind.

Since aerodynamic drag at this speed is about half the total resistance, it might be expected that the fuel savings would be roughly half the value of the wind-averaged drag reduction: 7.1 per cent and 13 per cent for the two configurations. This compares reasonably well with the figures of 6.2 per cent and 11.8 per cent fuel savings actually measured. All three vehicles were stripped to their baseline configuration and retested to check for differences in fuel consumption. In calculating the fuel savings, account was taken of the fuel consumption differences found. The measured differences between apparently identical models amounted to approximately 5 per cent, which indicates that good vehicle maintenance and tuning are of nearly the same order of importance as aerodynamic optimization.

During the above set of tests, the three vehicles were driven in a well-spaced convoy. In real-life operations, the driver would probably take some advantage from the improved performance of the drag-reduced vehicle to improve his speed on steep gradients, for instance where the speed is limited by engine power rather than by law. In this case, the fuel savings would be further reduced. Figure 5.28, which is also derived from the above source, shows how the fuel consumption would be affected if the vehicle speed were limited by the power available. A 240 kW (322 bhp) vehicle would be power-limited to around 103 km/h (65 mph). With the drag-reduction devices, this would rise to 113 km/h (71 mph). The fuel saving produced relative to the baseline vehicle would be 8 per cent instead of the 16 per cent reduction obtainable if the speed of 103 km/h were maintained.

Buckley estimated annual savings of around around 2500 US gallons per vehicle in the USA, based on an average of 160 000 km (100 000 mi) per year. Similar findings have been reported in the UK. Render[11] reported average fuel savings of between 9.7 per cent and 13.3 per cent (depending on the gearbox used) for a fleet of rigid-chassis trucks fitted with a drag-reduction kit. The baseline vehicle had a factory-supplied cab-roof fairing, and the streamlining kit produced drag coefficient reductions of 20 per cent at 0° yaw, rising to 25 per cent at 10°. The coefficients were measured using a 1/12 scale wind tunnel model. Assuming that aerodynamic drag represents about 50–60 per cent of the total drag, it can be seen that the fuel savings are roughly of the order expected, particularly when wind-averaging, Reynolds number and turbulence influences are considered. The baseline vehicle for the latter study already had a drag-reducing cab-roof fairing, so aerodynamic drag would have been a smaller proportion relative to rolling resistance than on older designs.

In summary, the above results and other full-scale studies generally give lower fuel savings than would be expected from simply looking at the reduction in zero yaw drag coefficient obtained in wind-tunnel tests. Correcting for the effects of the real average wind conditions produces more realistic estimates, although a further factor of 0.8–0.9 to take account of driving technique and other minor effects seems to be required.

It should be noted that buses are permitted to operate at speeds of 120 km/h (75 mph) or more in some countries, and at this speed the aerodynamic drag represents a much higher proportion of the overall resistance, so the potential savings on buses are even higher than for large trucks.

DESIGN REQUIREMENTS FOR LOW-DRAG COMMERCIAL VEHICLES

Despite the brick-like form of most commercial vehicles, their drag coefficient need be hardly any larger than for a domestic car. Rounding the front edges sufficiently produces a large drag reduction. Some rearward raking of the front face is also advantageous. In the case of articulated trucks, it is important that the tractor and trailer should as far as possible be matched to produce an integrated near continuous form, with the gap being sealed, and no discontinuity of line either in side or plan view.

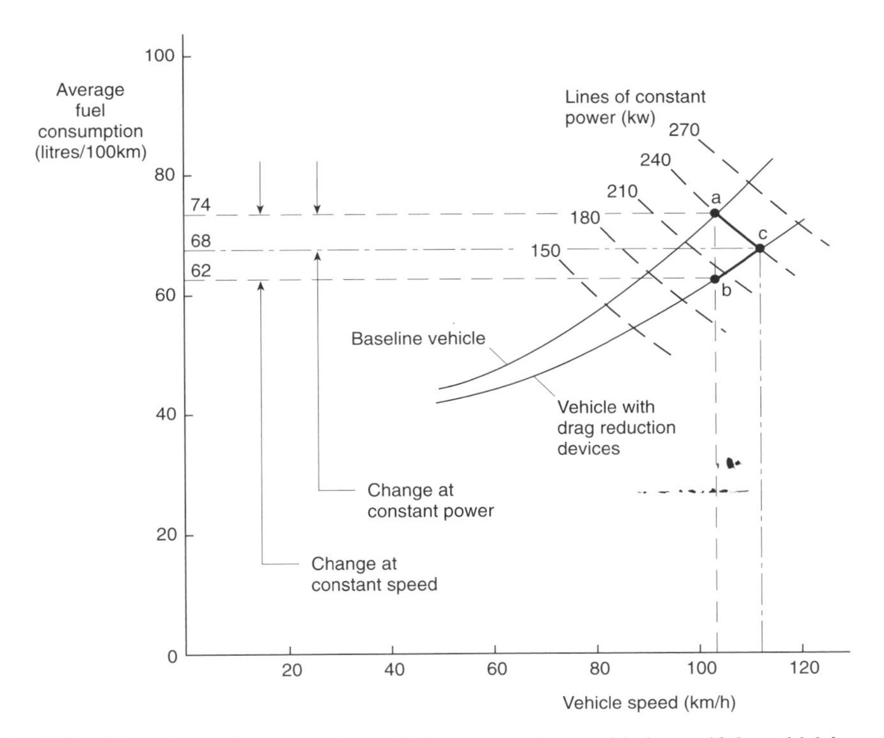

Fig. 5.28 How the maximum speed and fuel consumption would change if the vehicle's performance were limited by engine power rather than speed limits.
(*Based on data in Buckley*[1])

The economic advantages of streamlining are considerable even in urban driving, and if the vehicle is designed for low drag from the outset, there need be no penalty in terms of initial cost. Finally, it can be noted that styling commercial vehicles for aerodynamic advantage can produce visually attractive shapes that can help to enhance a company's image without detracting from other aspects of functionality or raising the cost.

REFERENCES

1. Buckley F. T., Marks C. H., and Walston W. N., A study of aerodynamic methods for improving fuel economy, US National Science Foundation, final report SIA 74 14843, University of Maryland, Dept. of Mech. Engineering, 1978.
2. Carr G. W., The aerodynamics of basic shapes of road vehicles, Part 1, Simple rectangular bodies, MIRA Report No 1982/2.
3. Dare-Bryan V., Winds of change – design development of commercial vehicle aerodynamics, *Proc. Vehicle Aerodynamics Conference*, Loughborough, 18–19 July 1994, RAeS, London 1994, pp. 20.1–10.
4. Garry K. P., Development of container mounted devices for reducing the aerodynamic drag of commercial vehicles, *Proc. 4th Colloquium on Industrial Aerodynamics*, 1–20 June 1980, pp. 175–90.
5. Gilhaus A., The influence of cab shape on air drag of trucks. *Proc. 4th Colloquium on Industrial Aerodynamics*, 1–20 June 1980, pp. 133–56.
6. Gilhaus A., The aerodynamic drag of buses, paper XI-3, *Proc. Colloque Construire avec le vent* (*Designing with the Wind*), 15–19 June 1981, CSTB Nantes.
7. Hoerner S. F., *Fluid Dynamic Drag*, Hoerner Fluid Dynamics, PO Box 342, Brick Town, N.J. 08723, USA, 1965.
8. Hucho W. H., *Aerodynamics of Road Vehicles*, Butterworth, London, 1987.
9. Koenig-Fachsenfeld R., Vom Zeppelin auf Radern bis zur K-form (Ein historischer überblick), *Proc. 4th Colloquium on Industrial Aerodynamics*, 1–20 June, 1980, pp. 1–16.
10. Palowski F. W., Wind resistance of automobiles, *SAE Journal*, vol. 27, 1930, pp. 5–14.
11. Render R. M., Drag reduction of commercial vehicles – experiment and practice, *Proc. Vehicle Aerodynamics Conference*, Loughborough, 18–19 July 1994, RAeS, London 1994, pp. 32.1–8.
12. Rouse and McNown, Cavitation pressure distribution of head forms, Iowa Univ. Eng. Bulletin No. 32, 1948.

CHAPTER 6

RACING CARS AND OTHER HIGH-PERFORMANCE VEHICLES

Racing car design might appear at first sight to be a less serious aspect of vehicle aerodynamics, but motor racing worldwide represents a large industry. Motor sport is used as a proving ground for the development of technologies that are subsequently applied to road-going vehicles, and it has inspired a great deal of aerodynamic research work.

The design of racing cars is currently strongly influenced by aerodynamic considerations. In order of relative importance, these may be listed as:

1. A high down force (negative lift) should be provided in order to increase adhesion to the road.
2. Down force between front and rear wheels, which affects the handling, particularly in cornering, should be balanced.
3. Drag should be minimized.
4. Adequate cooling air for engine, transmission, brakes, and driver should be provided.

Since the mid 1960s, the main interest in racing car aerodynamics has been in the generation of down force, but drag minimization is also important, and we will start with this topic. It will be convenient to illustrate the basic principles of low-drag racing car design by looking at some of the significant historical developments.

DRAG – THE HISTORICAL PERSPECTIVE

In the early days, most racing car designers paid little regard to aerodynamic drag. An example of this lack of concern is evidenced by the Star shown in Fig. 6.1. Competition cars of this early period were surprisingly fast; the similarly shaped Itala of 1903 was capable of 160 km/h (100 mph). Bearing in mind that above about 90 km/h (50 mph) aerodynamic forces begin to dominate, the lack of streamlining would have had a significant effect. The concept of streamlining was known at that time, and sometimes employed, but the practical

Fig. 6.1 Early racing cars such as this Star often showed no concession to aerodynamic considerations despite surprisingly high top speeds.

disadvantage was that any kind of panelling restricts access to mechanical components, and the early cars needed frequent wheel changes, or even running repairs. Another problem of streamlining was that as a consequence of following aeronautical practices, the designs often generated positive lift, which adversely affected handling and stability.

For the first three decades, racing car design was based largely on intuition and experience. A major advance occurred in the 1930s when the German (Nazi) government recognized that motor racing success would benefit the national image. The motor industry was encouraged to take racing seriously, and generous funding was organized. More importantly perhaps, academic and research institutions were drawn in to design the cars on a proper scientific basis. The resulting designs of Mercedes and Auto Union are now motoring legend. Wind-tunnel testing was used, and as shown in Fig. 6.2, some of the vehicles were well streamlined. The Auto Union is particularly interesting in that like most modern racing cars it had an aft engine layout. The resulting shape conforms quite closely to the ideal teardrop configuration.

After the war, Grand Prix racing cars had a moderately streamlined appearance (Fig. 6.3), but this probably came more from developing the pre-war forms than from any proper scientific approach. Racing sports cars such as the Jaguar D-type (Fig. 2.3) were derived from the open road-going cars, and despite the sleek appearance, had rather high drag coefficients, mainly due to the open cockpit. In the 1950s, Cooper developed the use of a rear- or mid-engined layout on the 500 c.c. formula, and by the 1960s had moved into Formula 1, where the layout rapidly became the accepted standard. Together with Lotus (Fig. 6.4) they

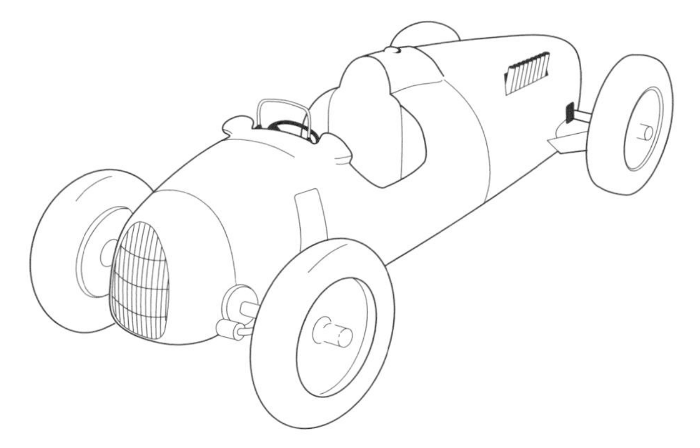

Fig. 6.2 The pre-war Auto Union
An approximation to the ideal low-drag teardrop profile. This advanced vehicle had a rear engine, all-round independent suspension and over 373 kW (500 bhp) of power. Its suspension geometry and weight distribution however gave rise to fearsome handling characteristics that only a few drivers could master.

Fig. 6.3 A post-war Maserati
A good basic shape, but with many excrescences and drag-producing features.

developed lighter vehicles with a small frontal area, which is of course just as important as a low drag coefficient. From that period onwards, serious attention has been paid to drag minimization in virtually all classes of racing car. With the introduction of wings in the mid sixties, however, aerodynamic attention began to be focused more on the provision of down force, as we describe later.

Fig. 6.4 An early rear-engined Lotus
The rear engine and reclined driving position permitted a very small frontal area and a well-streamlined form, except at the rear where access and exhaust system geometry took precedence.

DRAG MINIMIZATION

Major sources of drag on the early cars were the exposed suspension components and the open cockpit, features that are evident on the Napier Railton shown in Fig. 6.5. No attention at all was paid to the underside, or to parts that the eye could not see. By contrast, in modern designs, great care is taken to ensure that items

Fig. 6.5 The Napier Railton of the late 1930s
A great deal of parasite drag would result from the many exposed components, but with a 27 litre Y-configuration (three banks of four cylinders) aero engine there was no shortage of power.

such as springs and dampers are mounted out of the airstream. Components such as suspension struts that cannot be enclosed within a streamlined envelope are given a streamlined form (Fig. 6.6), and the underside of the vehicle is made as smooth and continuous as possible.

One of the problems of racing cars is that their internal layout often makes them incompatible with a low-drag teardrop form, which would require the widest and tallest component to be within the front third. With a front-engined layout, the driver's head needs to be above the engine, and will obviously be fairly well aft. The traditional solution was to locate the driver's head within a separate streamlined shape mounted on top of the basic body form. The same principle is still used on modern open-cockpit racing cars (see Fig. 6.7). By placing the engine behind the driver, the Auto Union (Fig. 6.2) went some way towards producing an arrangement that could be enclosed by a near teardrop shape.

Following the early lead of manufacturers such as Cooper, Lotus and Brabham, modern racing cars achieve a very small frontal area by adopting a highly reclined position for the driver, with the engine behind. The reclined driving position results in a relatively long nose to accommodate the driver's legs. On Grand Prix, Indianapolis and other open-wheeled cars, the nose widens gently in both plan and profile up to the head and shoulder position. The nose shape is often quite pointed in order to allow the largest possible area of uninterrupted flow over the front wing,

Fig. 6.6 Large suspension components such as springs and dampers are kept out of the airstream on this 1994 Williams. The slender suspension struts have a streamlined cross-section to minimize drag. Note also the floor panel, which is extended forward to form a splitter plate under the nose.

Fig. 6.7 Williams Formula 1 of 1994 showing the nicely streamlined integration of the head fairing and ram-air intake box with the engine cover.

which is normally mounted either on, or now more commonly under, the nose. Although the classic teardrop shape cannot be used, the narrow form means that it is possible to avoid strongly adverse pressure gradients over most of the body length, and the resulting $C_D A$ is low. It should be remembered that minimizing the area is just as important as lowering the drag coefficient.

DRAG DUE TO WHEELS

Some racing car classes such as Formula 1 specifically require that the wheels should be fully exposed. Exposed wheels produce a significant amount of drag because the frontal area of each tyre is a little under half the frontal area of the bodywork, and the drag coefficient of an exposed wheel is high. Measurements of the drag coefficient for an open-wheeled racing car *without wings* vary from around 0.7 for an idealized wingless model of a 1970s Formula 1 (Stapleford and Carr[14]) to around 0.8 for a late 1980s vehicle (extrapolated from data presented in Benzing[3]). Of this figure, about 70 per cent is attributable to the exposed tyres. Note that with wings fitted, the overall drag coefficient will be around 1 or more, the exact figure depending on the wing arrangement, which in turn depends on the characteristics of the circuit.

Despite the low overall frontal area, the $C_D A$ value of an open-wheeled car will be high. Table 6.1 shows a comparison between a typical Formula 1 car and a Group C sports car of the late 1980s (Fig. 6.8). At around 1.2 m^2 (13 ft^2) the Formula 1 car with a wing has a $C_D A$ value comparable with a small delivery van.

Table 6.1 Typical C_D and C_DA values for closed- and open-wheeled racing cars.

Car	Without wing		With wing, medium circuit	
	C_D	C_DA (m^2)	C_D	C_DA (m^2)
Formula 1	0.8	1.02	0.83	1.2
Group C sports	0.24	0.38	0.35	0.55

Fig. 6.8 The highly successful Spice Tiga Group C racing sports car of 1983
The all-enveloping bodywork produced a low drag coefficient: 0.24 without the rear wing.

With exposed wheels, the flows over the front and rear pair interact with each other, the greater effect being on the rear pair, which are in the wake of the front pair. It is therefore important to measure the overall drag from a complete arrangement of four wheels. Stapleford and Carr[14] measured the drag coefficient of typical exposed rotating racing car wheels in contact with the ground. With four wheels arranged as on a typical racing car, the drag coefficient based on the projected frontal area (i.e. the frontal area of a pair of wheels) was approximately 1. Benzing[3] quotes a consistent value of 1.051. For two wheels (on a single axle) the drag coefficient value was found to be 0.6 based on the same frontal area.

Table 6.2, which is based on a combination of the author's own data and published data including that of Benzing[3], shows the breakdown in drag contributions on a typical Formula 1 car with a wing set for a medium fast circuit. Lift forces are dealt with later in this chapter.

Table 6.2 Drag breakdown for a Formula 1 car on a medium fast circuit. Areas are projected frontal areas.

Contribution	Area A (m^2)	C_D	$C_D A$ (m^2)	C_L	$C_L A$ (m^2)
Wings	0.15	1.0	0.15	} −3.5	−3.0
Body	0.7	0.6	0.42		
Tyres	0.53	1.0	0.53	0.6	0.32
Overall less wing	1.23	0.77	0.95	—	—
Overall with wing	1.38	0.8	1.1	−1.94	−2.68

(Based on data from a variety of sources, including Benzing[3])

INCORPORATING THE COOLING SYSTEM WITHIN A LOW-DRAG ENVELOPE

A fundamental feature of the petrol/gasoline engine is that the maximum efficiency is only around 33 per cent, so that for every kilowatt of power produced, there is at least two kilowatts of waste heat to dissipate. Racing engines can produce 560 kW (750 bhp) or more, and although the average power will be much less than this, the cooling system needs to dissipate around 200–300 kW. To get some idea what this means, imagine the output from over two hundred electric fires! Trying to locate the necessarily large radiator within the same streamlined envelope as the engine and driver is extremely difficult with open-wheeled racing cars. The current solution is to place the driver and engine in a narrow streamlined envelope, with the radiators and fuel tanks incorporated into a pair of side pods (see Fig. 6.7). The resulting wide overall shape is useful for producing down force by ground effect, as described later, and the pods are now made as wide as the racing regulations permit. A similar internal layout is often used for racing sports cars, but the totally enveloping bodywork results in a much more obviously streamlined flattened form (Fig. 6.8). Apart from their effect on the external aerodynamics, the radiator systems can produce a great deal of internal drag. The design and optimization of cooling systems for low drag are described in some detail in Chapter 7.

Current open-wheeled racing cars do not look particularly streamlined, but this is because it is necessary to integrate low-drag principles within the constraints imposed by the down-force-producing systems. In order to understand more recent developments, it is therefore necessary to look at down force, and the methods that are employed to create it.

AERODYNAMIC DOWN FORCE

In the 1960s the use of soft rubber compounds and wider tyres, pioneered particularly by Lotus, demonstrated that good road adhesion and hence cornering ability was just as important as raw engine power in producing low lap times. The tyre width factor came as something of a surprise, because in simple school experiments on sliding friction between hard surfaces, the friction resistance force is found to be independent of the contact area. It came as a similar surprise to find that the friction or slip resistance force could be greater than the contact force between the two surfaces, apparently giving a coefficient of friction greater than 1. As was explained in Chapter 3, however, for elastic compounds such as rubber in rolling contact with a rough surface, simple sliding friction principles are inappropriate. The tyres do not actually slide relative to the road under normal circumstances; the rubber merely distorts elastically. As described in Chapter 3, the maximum lateral (sideways) 'grip' force on a tyre is related to the down load by

$$\text{Maximum lateral force} = k_c(\text{max}) \times \text{down load}$$

where $k_c(\text{max})$ is the maximum lateral adhesion or cornering coefficient, which can be around 1.4 for racing 'slick' tyres.

The desire to further increase the tyre adhesion in cornering led to a major revolution in racing car design: the introduction of inverted wings, which produce negative lift or 'down force'. Since the tyre lateral adhesion is roughly proportional to the down load or contact force between road and tyre, adding aerodynamic down force to the weight component will obviously improve the adhesion.

If the suspension set-up and the weight and aerodynamic down force distribution were all perfectly matched, the maximum cornering force available would approach the value obtainable on an isolated tyre, and this force could be equated to the centripetal force:

$$(W/g) \times (V^2/r) \rightarrow k_c(\text{max})(W + L_d) \qquad [6.1]$$

where V is the speed

L_d is the aerodynamic down force

W is the vehicle weight

g is the gravity constant

Using the above expression it is possible to calculate the ideal effect of aerodynamic down force on cornering speed for a given radius, as illustrated in Fig. 6.9, where a corner of 40 m radius (131 ft) has been chosen. Note that this is a hypothetical maximum limiting case, and a real vehicle would not be able to achieve such a cornering performance because it is not in practice possible to get all four wheels to achieve the theoretical maximum side force simultaneously; the relative importance of aerodynamic down force is however illustrated.

As described in Chapter 3, simply increasing the weight does not improve the cornering ability, as although the down force and hence grip improves, the increase in mass means that the centripetal force required increases by an equal amount.

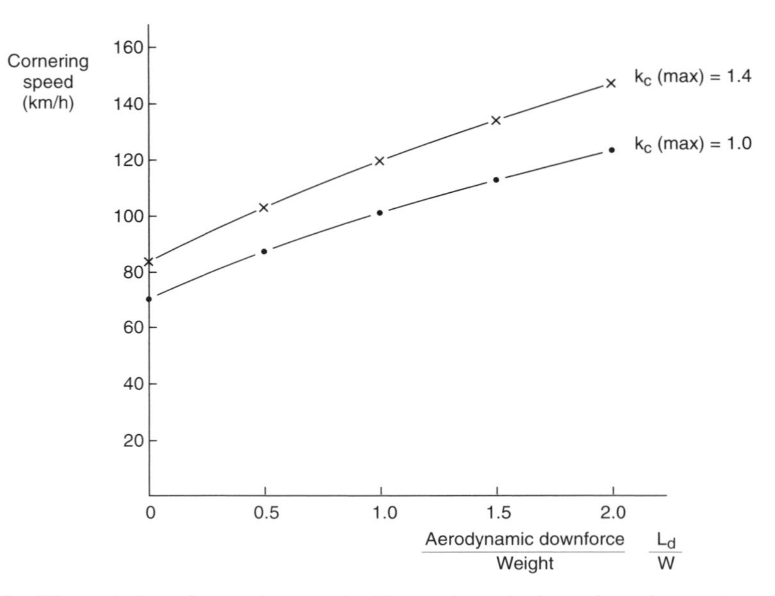

Fig. 6.9 The variation of cornering speed with aerodynamic down force for a radius of 40 m, and different values of maximum effective tyre cornering coefficient k_c(max).

THE PHYSICAL AND MECHANICAL STRESSES OF ENHANCED CORNERING ABILITY

An important factor to consider when trying to increase the cornering ability of a car is the loads that this will impose on both the car and the driver. The limiting case of the expression [6.1] above can be rearranged to give the hypothetical maximum centripetal acceleration.

$$\text{centripetal acceleration} = V^2/r \rightarrow k_c g\,(1 + L_d/W) \qquad [6.2]$$

A Formula 1 car on a medium fast circuit in the early 1990s would typically have a reference area of about 1.4 m^2 (15 ft^2), and produce a C_L value of around -1.95. The aerodynamic down force L_d that this would produce at 300 km/h (188 mph or 83 m/s) can be calculated as 11 529 N, which is around twice the weight of a typical Formula 1 car.

$$L_d = 2W$$

Putting this into equation [6.2] and taking a k_c(max) value of 1.4 for soft slick racing tyres, it is found that in the limiting case of perfect set-up and balance, the centripetal acceleration approaches 4.2g.

The physical demands that this level of acceleration imposes on the driver is severe; racing drivers have to be fit, and need strong neck muscles. An interesting thought is that with the down force greater than the car weight, it would be theoretically possible to drive upside down along the roof of the Monaco tunnel!

High down force is not limited to pure racing cars; modified saloon cars with air dams and spoilers or wings can provide significant amounts of down force. Gerhardt *et al.*[6] reported lateral accelerations of up to 1.6*g* for a modified racing Ford Capri in 1980.

DOWN FORCE AND ITS INFLUENCE ON PERFORMANCE

In addition to enhancing the cornering ability, aerodynamic down force allows the tyres to transmit a greater thrust force without wheelspin (as described in Chapter 3), and hence the acceleration will be increased. Without aerodynamic down force, high-performance racing cars have sufficient power to produce wheelspin up to more than 160 km/h (100 mph). Taking typical Formula 1 car data as used in the previous example, it is possible by numerical integration to calculate the influence of the aerodynamic down force on the vehicle's acceleration. The calculation is quite complicated because the inertia effects on acceleration or 'weight transfer' also increase the down force on the rear wheels. Table 6.3 compares the performance with and without a lift coefficient of −1.96, and running on slick tyres with $k_s(\max) = 1.4$, and rear-wheel drive. It has been assumed rather unrealistically that the aerodynamic down force is distributed equally between the front and rear wheels, and that the tractive force is limited by adhesion rather than power at all speeds. The centre of gravity is assumed to be located mid-way between the two wheels and 0.5 m (1.6 ft) above the road. Although the down force does not make a large difference to the acceleration time from rest, there is a considerable improvement in acceleration at 160 km/h (100 mph).

Table 6.3 The effect of aerodynamic down force on the acceleration of a racing car in the adhesion-limited case.

Measurement	With down force ($C_L = -1.96$)	With no down force
Time from rest to 44 m/s (160 km/h)	5 s	6.06 s
Final rate of acceleration (i.e. at 44 m/s)	10.02 m/s^2	5.52 m/s^2
Amount of power transmitted at 44 m/s	353 kW	229 kW

INFLUENCE ON BRAKING

Deceleration forces are affected in the same way as thrust forces. The brakes are capable of locking the wheels at almost any speed, and deceleration forces like thrust forces are limited by the adhesion between the wheels and the road. With aerodynamic down force equal to twice the weight of the car, and a tyre maximum slip coefficient k_s of 1.4, the initial braking deceleration will be around $4g$. To the novice driver, this would feel like hitting an obstacle.

In addition to performance improvements, aerodynamic down force tends to improve the directional stability of the vehicle, as described in Chapter 10.

THE PRODUCTION OF DOWN FORCE - WINGS

The aerodynamic down force revolution started at the end of 1965 with the appearance of the Chaparral 2C racing sports car (Fig. 6.10). This vehicle employed a large rear wing to produce down force. In 1966 a development, the Chaparral 2D, performed extremely well, and achieved a victory at the Nurburgring. The car displayed some advanced aerodynamic features, such as controllable wing incidence (not now permitted under racing rules). Rather surprisingly, despite this success, wings were not applied on Formula 1 cars until mid 1968, when the Ferrari V12 3000 appeared with a wing, and immediately demonstrated its advantages.

There were some earlier historical examples of vehicles with inverted wings, such as the Opel rocket-propelled car of 1928 (Fig. 6.11), and the Mercedes Benz T80 of 1939. Both these vehicles were used for record breaking rather than racing, and the wings, which were mounted amidships, were intended primarily to prevent the car from taking off, or at least from losing adhesion at high speed. A precedent for the use of wings on track racing cars was provided by the Swiss

Fig. 6.10 The Chaparral sports racing car, which introduced the concept of using inverted wings to increase road adhesion.

Fig. 6.11 An early example of a winged high-performance vehicle, the Opel Rak.2 rocket-propelled car, designed for speed record attempts. Note that although the wing has sufficient negative incidence to provide down force, it is cambered as for positive lift, suggesting that the designer's knowledge of aerodynamics was rather limited. A mixture of gun-cotton and potassium chlorate propelled the vehicle to 100 km/h (62 mph) in 6 s, not impressive by modern racing car standards.

engineer Michael May, who mounted a wing on a Porsche in the mid 1950s. However, this was immediately banned as dangerous, and had no impact on racing car design.

A major disadvantage of a fixed wing is that it produces a high drag, and therefore, when wings were first introduced to Formula 1, they were (as on the Chaparral) provided with a mechanism for controlling the incidence angle. On the straight, the wing could be feathered to a zero incidence to minimize drag, but on braking and entering a corner, the incidence was increased negatively to provide down force. Front wings were soon added, and as a further refinement, Brabham in Formula 1 and Porsche in sports car racing introduced split wing surfaces with independently variable incidence. These enabled a greater down force to be applied to the inside wheel to offset the effects of the rolling moments. In the early experiments of 1968, Brabham and Ferrari mounted the wing on the body, but on the Lotus, the wing was mounted directly on to the unsprung wheel assembly so that the down force would not push the bodywork down into contact with the road. In order to take the wing out of the slipstream, Lotus pioneered the use of very high mounted wings, as illustrated in Fig. 6.12: an arrangement that was taken up by a number of other competitors.

The combination of high mounting, variable incidence mechanisms and direct connection to the unsprung wheel assembly resulted in a flimsy and vulnerable structure that was susceptible to accidental damage. Following serious accidents at the Barcelona Grand Prix in 1969 stringent regulations governing the use of wings were introduced, and these are still in force. They are essentially:

1. The maximum height, width and locations are controlled.
2. Aerodynamic devices have to be of fixed geometry (no variable incidence).
3. They have to be rigidly attached to the bodywork.

The last two regulations have caused significant problems. The fixed incidence means that a compromise has to be achieved between the improved cornering due to high down force, and the reduced speed resulting from the high trailing vortex

Fig. 6.12 Bruce McLaren's Formula 1 M7-C was one of a number of cars which appeared briefly with a very high wing mounting position. This arrangement was quickly banned during 1969 because of a number of accidents.

drag. The optimum compromise differs according to the nature of the circuit, and considerable effort in practice sessions goes into trying to achieve the right balance. The requirement for attachment to the bodywork means that the ground clearance varies with the down load. The only solution is to severely increase the stiffness of the suspension springing, which unfortunately produces a jarring uncomfortable ride for the drivers, and can cause spinal injuries. Over the years, the racing regulations in all classes have been progressively tightened, with regulators and constructors playing cat and mouse.

Before going on to a description of the features of car wing design, it will be useful to look at another milestone in the development of racing car aerodynamics, the introduction of the ground effect to increase down force.

GROUND EFFECT – HISTORICAL DEVELOPMENT

At the end of 1976, the Lotus 78 appeared featuring what was in effect a pair of inverted stub wings, formed by shaping the side pods on each side of the cockpit. The basic arrangement was similar to that of the later Lotus 79 shown in Fig. 6.13. The wings were enclosed by side plates which extended down to give a rubbing contact with the ground. Being in such close proximity to the ground, the wings were much more effective than a high mounted or isolated wing, and more importantly, produced much less associated trailing vortex drag. The underside of the wing section formed a convergent–divergent 'venturi' channel with the road, generating a low pressure, effectively sucking the vehicle down. The side panels prevented any inflow of air from the sides, as this would have tended to destroy the

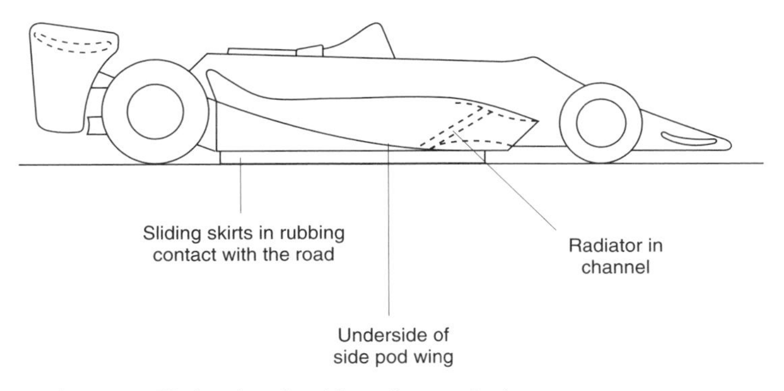

Fig. 6.13 The Lotus 79 showing the side pod venturi wing arrangement.

Fig. 6.14 The Brabham BT-46 B 'vacuum cleaner' of 1978
The large engine-driven fan was used to reduce the underbody pressure, and hence improve down force. It was almost immediately banned.

suction. They were held in rubbing contact with the ground by springs. A more detailed explanation of how the underbody shape can be used to generate down force, and the reasons for the low drag penalty are given later in this chapter. The Lotus 79 enabled the team to dominate the 1978 season, and the ground effect idea and design features were taken up by rival teams.

A more extreme form of ground effect was the use of engine-driven fans to enhance the underbody suction, a technique explored by both Chaparral and Cooper but quickly banned. Figure 6.14 shows the Cooper BT-46 B 'vacuum cleaner' with the fan clearly visible. Some idea of the importance of ground effect down force may be gleaned from the fact that during the period of introduction, 1976 to 1982, lap times at Siverstone fell by 15 per cent, and at Long Beach by 20 per cent.

Unfortunately, as with the early wings, the strong underbody ground effect proved to be potentially dangerous. Any sudden reduction in down force produced by the vehicle hitting a bump or kerb when cornering on the limit, or simply coming into the wake of another car, would cause the vehicle to leave the track at very high speed (see Howell[7]). Consequently, a series of safety measures were introduced; the sliding skirts were banned, and a regulation was introduced which specified that a certain percentage of the underside had to be flat. This latter rule was found to be less effective than had been expected, because designers soon realized that it was unnecessary to have an inverted aerofoil shape under the car; low pressure on the underside could be created as long as a diffuser, a section of gradually increasing area, was created at the rear end of the underside (as described later).

The effect of having to remove the sliding skirts was partly overcome by use of an extremely small ground clearance. Following the deaths of Roland Ratzenberger and the reigning Formula 1 champion, Ayrton Senna, in 1994, further restrictions aimed at reducing cornering speeds were introduced. These included measures to limit ground effect by the introduction of a step in the underside cross-section, which effectively controlled the ground clearance.

The large amounts of down force generated by ground effect mean that less is required from the wings, and for some classes, and fast tracks, wings are used mainly as down force trimming devices. The balance between wing and underbody down force changes with the rules; in 1982 when large amounts of underbody down force were permitted, it provided some 80 per cent of the total.

Despite concerns over safety, and ever tighter regulations, it seems likely that aerodynamic down force will continue to be used on many classes of racing car for the foreseeable future. The design features of both wings and ground-effect underbodies are therefore described in some detail below.

RACING CAR WING DESIGN

Although underbody suction in important in generating down force, wings still play a part, both as a major contributor to the overall down force, and as a readily adjustable trimming device used to control the distribution of vertical reaction between front and rear wheels. Wing-generated down force is also safer than underbody suction, as it is less sensitive to the pitch attitude and ride height of the car.

Some relevant definitions and terms used in wing aerodynamics are shown in Fig. 6.15. Note that two different definitions of C_L are commonly used. When isolated wings are being described, the aeronautical definition is used, namely:

$$\text{wing } C_L = \text{wing lift} / \tfrac{1}{2}\rho V^2 \times (\text{wing plan area})$$

but when dealing with the overall lift effect on a car, the definition used is:

$$\text{overall } C_L = \text{car lift} / \tfrac{1}{2}\rho V^2 \times (\text{car projected frontal area})$$

Down force is of course just negative lift, and C_L values for car wings are therefore invariably negative.

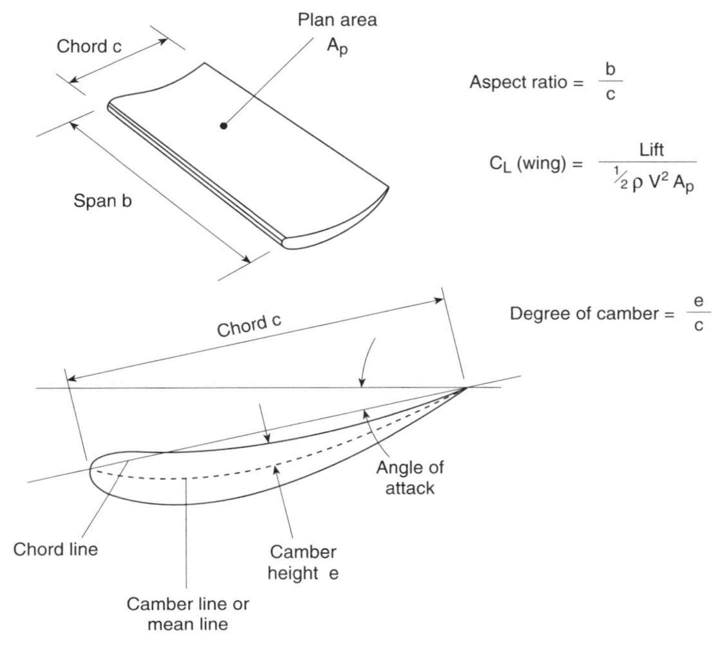

Fig. 6.15 Some key aerofoil section definitions.

DIFFERENCES BETWEEN CAR AND AIRCRAFT REQUIREMENTS

The constraints placed on car wings by racing regulations and practical size limitations mean that they are optimized in a completely different way from those of aircraft. For a racing car, regulations severely limit the size of the wing, and for Formula 1 cars on a twisting circuit, the requirement is for a wing that will produce a high down force per square metre of area. This means that a high lift coefficient is required. Aircraft wings on the other hand are designed primarily to give a large lift to drag ratio for high efficiency at their cruising speed. This favours wing sections with only very small degree of camber (for definition see Fig. 6.15) operating with a lift coefficient in cruise of usually less than 0.5. To find an aeronautical precedent for high C_L wings it is necessary to look at the special arrangements of flaps and slats adopted by aircraft when landing. Even these designs are, however, not ideal for translation to cars, as in aircraft the shape is restricted by the need for the various elements to retract into a cruise-optimized aerofoil shape.

Not all classes of racing car use the same high C_L values as Formula 1, and in many categories, the compromise between straight-line speed and cornering ability requires a more modest down force. Indy cars when set up to race on fast oval circuits require little down force and a low drag. As was shown in Chapter 2, the

Fig. 6.16 Formula Renault cars with restricted power cannot employ too much wing down force because of the drag penalty. Simple single-element aerofoils may be seen on this vehicle.

generation of lift usually results in the production of trailing vortex drag, so for this case, the type of wing sections used for aircraft might be appropriate, as they are optimized for low drag at a moderate lift coefficient. Group C sports cars are permitted much wider wings (Fig. 6.8) than are allowed on Formula 1, and therefore they also do not always require very high C_L values. In some of the minor classes (Fig. 6.16), the engine power is quite limited, so the trailing vortex drag penalty imposes limits on the amount of down force that can be generated. From this, it may be seen that to cover the full range of racing car classes, it is necessary to look at a variety of wing designs optimized for different C_L values.

LOW C_L WING SECTIONS - SELECTING OR DESIGNING THE CORRECT AEROFOIL

One of the most common mistakes in vehicle wing design is to think in terms of the characteristics of two-dimensional wing sections operating in an undisturbed flow. Car wings are generally of low aspect ratio (span/chord ratio) and operate either very close to the ground (at the front), or in close proximity to the bodywork at the rear of the vehicle, where neither the flow speed nor the pressure are at their free-stream values. High-performance car wing sections have to be designed with this in mind, and cannot be simply selected off-the-peg from standard profiles designed for aeronautical applications. These interference effects are described in more detail later. Despite such restrictions, the principles of aerofoil design as employed for aircraft at least provide a good starting point.

Where the wing C_L requirement is low (less than about 0.5), a single-element aerofoil with a small degree of camber can be used. Many of the more modern aircraft aerofoil sections are designed to encourage low-drag laminar flow over a relatively large proportion of the suction surface (the underside on a car down force aerofoil, corresponding to the upper surface on an upward-lifting aircraft wing). Figure 6.17 shows typical pressure distributions for two different aircraft aerofoil sections. It will be seen that for the newer section, the suction surface has a large portion with nearly uniform low pressure, and a slightly favourable pressure gradient over most of the front half. This is followed by a short section of adverse gradient. The larger proportion of favourable pressure gradient on the newer section helps to delay transition from laminar to turbulent flow, so this section will encourage a significant length of low-drag laminar boundary layer, giving a low surface friction drag. The friction drag on a car wing is, however, very small in terms of the total drag, and this factor is of less importance than on aircraft.

Aerofoils are usually designed as families with the same basic shape, but with variations in thickness, camber and the position of maximum thickness. Figure 6.18 shows typical C_L/C_D curves for two of the NACA 6-series family of aerofoils. (NACA, the National Advisory Committee on Aviation, was the forerunner of NASA.) The lift coefficient can be varied by changing the angle of incidence, and it will be seen that for a small range of incidences, there is a dip or 'bucket' in the drag curves. Within this range, the aerofoil operates with a large proportion of laminar boundary layer flow. Not all aerofoils are designed to produce a laminar 'bucket', as it is recognized that the required standard of surface finish, cleanliness

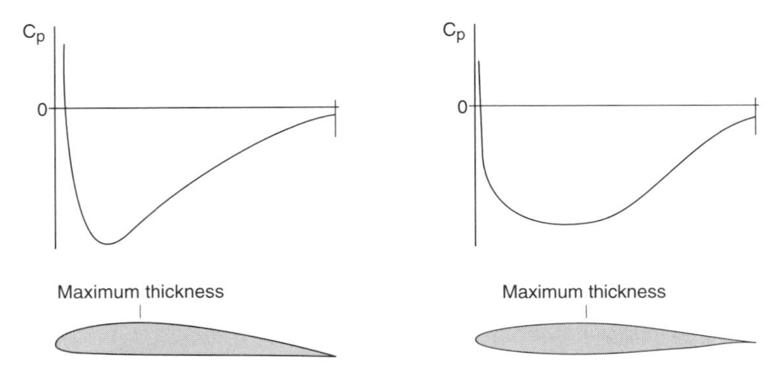

Fig. 6.17 Pressure distribution on older and more modern aircraft aerofoils. The high peak near the front of the older section meant that it only produced a short length of low-drag laminar boundary layer. Such aerofoils can have an advantage for cars when operated at high lift coefficient (high incidence), because the adverse pressure gradient is less severe, helping to maintain attached flow. At low lift coefficients, the modern aircraft section has a better lift to drag ratio.

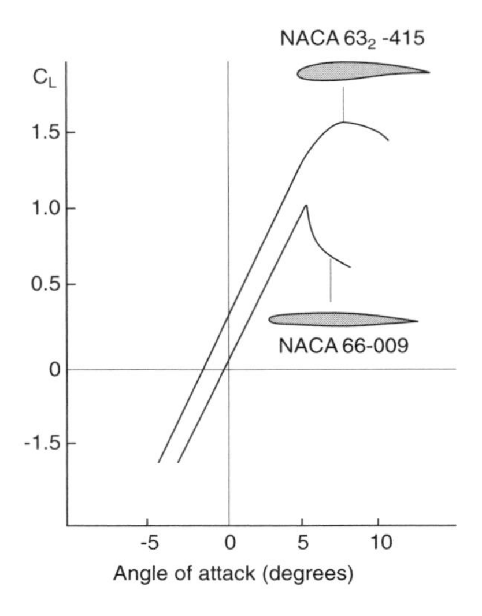

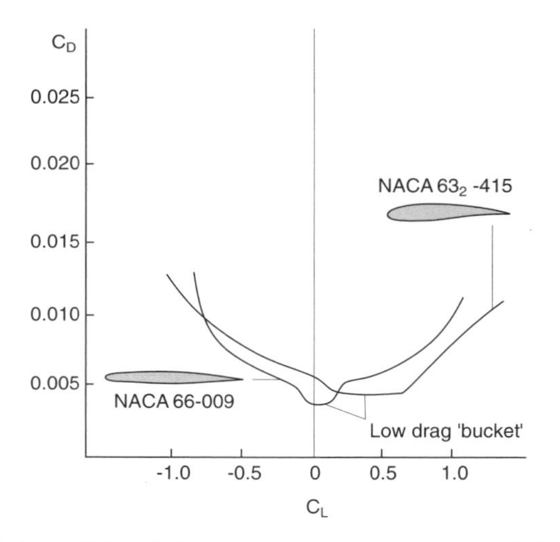

Fig. 6.18 The variation of lift and drag with angle of attack for two different aerofoils of the NACA 6-series family. The thinner aerofoil produces a lower minimum drag, but the operating range is much smaller.

Note that these lift and drag coefficient values are based on the plan area of the wing, whereas overall vehicle lift coefficients are usually based on the projected frontal area. To change the above coefficient values to ones based on car frontal area, it is necessary to multiply by (wing area/car frontal area).

and production accuracy is rarely achieved, but all sections have a limited range of lift coefficients for which the drag coefficient is at a low value.

The designation numbers of the aerofoils usually give information about the operating characteristics in coded form. Abbott and von Doenhoff[1] describe the designation codes of many of the NACA and other families of aerofoils.

Although the wings have to be fixed at one incidence when being raced, the incidence (and consequently the lift coefficient) is set up to suit the circuit and adjusted during practice sessions. It is therefore useful to have one aerofoil that produces low drag over a reasonable range of lift coefficient values. From Fig. 6.18 it can be seen that the thin aerofoil with the position of maximum thickness near the middle produces the lower minimum drag, but this is confined to a very narrow range of operating C_L values, and the thicker aerofoil would be a more practical choice.

There are many standard aerofoils designed for aviation purposes, but they are generally only useful for applications where a moderate down force is required. The influences of wing planform and ground or bodywork proximity also mean that significantly modified shapes may be required, as described in more detail later. Further information on aircraft aerofoils is given in Smith.[13]

WING SECTIONS FOR MODERATELY HIGH C_L

To design a single-element aerofoil that will generate a high C_L, it is necessary to produce a shape which avoids severely adverse pressure gradients (air flowing from a low pressure area to a higher one). An adverse pressure gradient will cause the boundary layer to thicken rapidly, resulting in high drag. More importantly, if the gradient is too severe, it will produce flow separation, with a loss in lift, and even greater drag: the well-known stall condition on an aircraft. The most critical region of adverse pressure gradient occurs on the low-pressure or 'suction' surface, which is the lower surface on a car wing designed for down force. The onset of separation is dependent not only on the severity of the adverse pressure gradient, but also on its length; the greater the length, the thicker will be the boundary layer and the greater its tendency to separate. Figure 6.19 shows two different suction-surface pressure distributions designed to give the highest possible lift without producing separation. Curve (a) is produced by having the maximum thickness well forward. The region of adverse gradient (pressure rising back towards local atmospheric) is long, allowing a very low pressure at the front, but the lift is limited by the fact that this low pressure acts over a relatively small area. Curve (b) is produced by an aerofoil with the point of maximum thickness much further aft. The region of adverse pressure gradient is shorter, which allows it to be steeper, but the lift is limited by the relatively poor maximum suction. Somewhere between the two extremes there lies an optimum compromise giving the highest lift coefficient.

Curves of this general form have been derived theoretically by Liebeck.[10] A further description is given in a very informative paper by Katz.[9] The aerofoil shapes necessary to produce these pressure distributions can be derived using a variety of computational fluid dynamics (CFD) methods. The resulting optimum

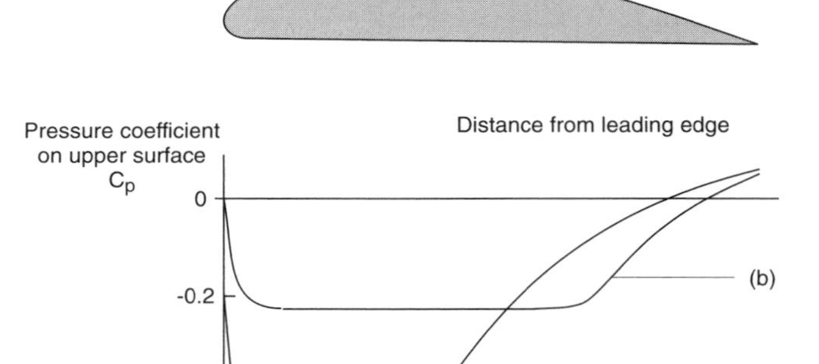

Fig. 6.19 Suction-surface pressure distributions on an aerofoil with a region of nearly constant low pressure, designed to give the maximum possible lift without producing flow separation. The highest lift coefficient will be given by a version somewhere between (a) and (b), as the best compromise between length and strength of the low-pressure area. Computational techniques can be used to generate a physical aerofoil shape that will produce the optimum pressure distribution.
(*After Katz,*[9] *courtesy of Prof. J. Katz*)

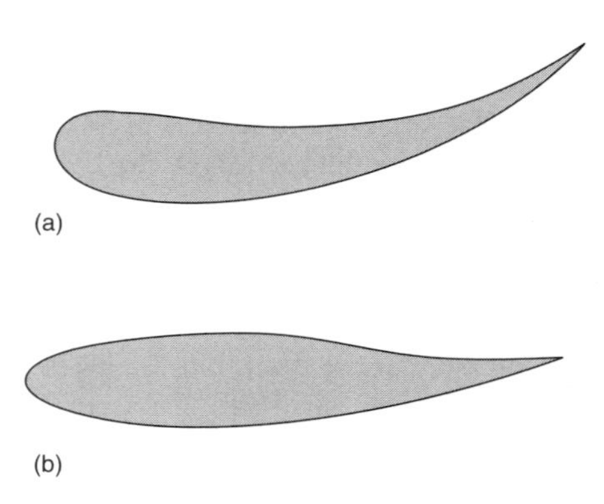

Fig. 6.20 (a) An aerofoil shape optimized for car use, (similar to one of a family of aerofoils given in Benzing[3]) compared with (b) a low-drag aircraft aerofoil, the NASA LS(1)-0417 or GAW(1) (see McGhee and Beasley[11]) optimized for low drag at moderate lift coefficient. Note how the point of maximum thickness is further forward on the car aerofoil, and the camber is much greater. The car aerofoil is optimized for a much larger lift coefficient, and would give more drag than the aircraft section. The aircraft aerofoil is of course normally used the other way up.

shape has the point of maximum thickness further forward than on most modern high-efficiency aircraft aerofoils, and has a higher degree of camber, as may be seen from Fig. 6.20.

Higher lift coefficients can be generated by artificially turbulating the boundary layer in the adverse region by using a rough surface, or even vortex-generating vanes, but this will be at the expense of an increase in drag.

VERY HIGH C_L WINGS

The methods of producing very high lift follow the example of aircraft in landing configuration, and in particular, aircraft designed for very short take-off and landing (VSTOL).

Flow separation due to adverse pressure gradient imposes a limit on the lift coefficient that can be obtained with a single-element cambered aerofoil. Separation is essentially caused by the lack of available energy in the boundary layer (lost through surface friction), and one way to overcome this problem is to split the aerofoil into a number of elements, as shown in Fig. 6.21. At each split, air flowing through the gap forms a new thin high-energy boundary layer on the following element. By using several elements, very high lift coefficients can be generated. As long ago as 1921, Sir Frederick Handley Page and Dr G. V. Lachmann produced C_L values as high as 3.9 using multi-element aerofoils. With modern design

Fig. 6.21 Multiple aerofoils with multiple elements and end-plates on the Ferrari 412 T1B of 1994. The upper wing has three elements or flaps. Note the downward cant or anhedral of the lower wings.

Fig. 6.22 Multi-element front wing with end-plates.

methods, much higher values have been obtained. Multi-element 'slotted flap' aerofoils are commonly used on Formula 1 and Indy cars both for the front and rear wings (Figs. 6.21 and 6.22).

THE DESIGN OF MULTI-ELEMENT HIGH-LIFT SECTIONS

The design of multi-element wings is largely a case of trial and error, guided by experience and an understanding of air flow behaviour. The location of one element in relationship to the next and the gap dimensions have to be optimized. Computational fluid dynamics (CFD) can be very helpful in the initial stages, but the complex interactions between the boundary layer and the wake of the preceding element make this quite a difficult problem for computational methods, and once again these have to be backed up by wind-tunnel verification and experimentation. The basic principle of avoiding highly adverse pressure gradients applies.

MULTIPLE WINGS

Whatever one wing can produce in terms of down force, two wings will produce more, though not usually twice as much, since the air flows will interact. If two aerofoils in a biplane arrangement are brought close together, the lift coefficient of both of them will drop, as shown in Fig. 6.23. This lift reduction can be understood by imagining what will happen when the two aerofoils are nearly touching. The lower surface of the upper aerofoil and the upper surface of the lower aerofoil will be at nearly the same pressure, so the resulting forces (which are in opposite directions) will cancel out, and the lift of the combination will be almost entirely attributable to

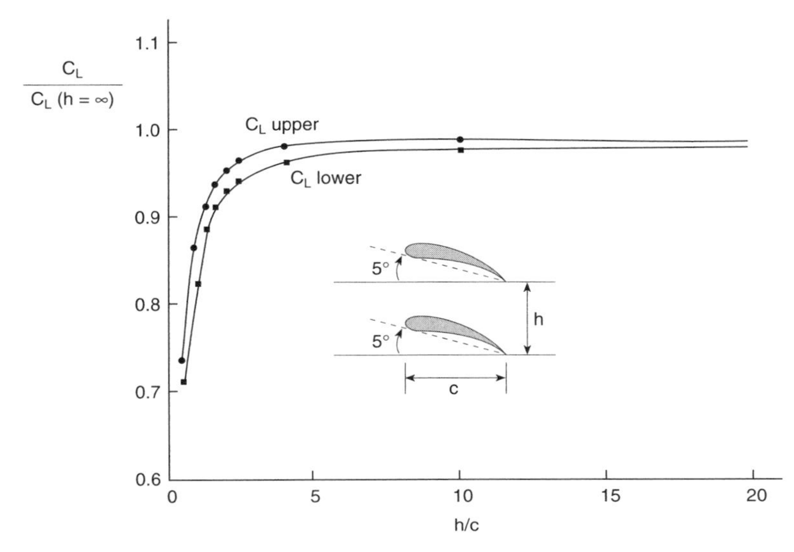

Fig. 6.23 The effect of changing the separation distance between two aerofoils mounted in a biplane arrangement.
(*After Katz,*[8,9] *courtesy of Prof. J. Katz*)

the pressure distributions on the outer surfaces. In the extreme case, the resulting lift will thus be similar to that produced by a single aerofoil.

Although the C_L of each individual wing may be reduced, the overall lift of two or more aerofoils is still greater than that of a single isolated aerofoil. Multiple wing arrangements are therefore useful when racing regulations restrict the overall dimension of the wing to such an extent that a single aerofoil cannot generate enough lift. Fig. 6.21 shows an extreme example of the multi-plane arrangement on a Ferrari Formula 1 car.

Figure 6.24 shows the effect of increasing the number of aerofoils within a given height limit. The lift coefficient increases with the number of aerofoils, but less than linearly, and the lift/drag ratio falls. It should be noted that although the interaction between wings may reduce C_L at a given incidence, it can help maintain attachment, permitting a higher incidence to be used, so the maximum attainable lift need not necessarily be reduced. There will, however, be a drag penalty.

Cars wings do not suffer the same structural and weight constraints as those of aircraft, but the problem of any complex arrangement is that it takes longer to optimize and set up than a simple one.

From the early 1990s it has become the practice for the secondary wing on some cars to be canted down towards the outside (Fig. 6.21). On aircraft this anhedral effect is introduced for reasons of stability, but on Formula 1 cars its purpose is to take the wing into a region of flow where it can be most effective, either by being less affected by the wake of forward components or because of a beneficial influence on the underbody venturi pressure, as discussed later.

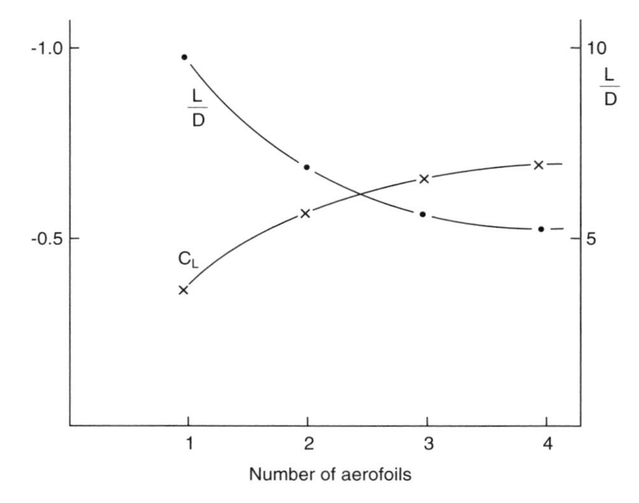

Fig. 6.24 The effect of using multiple wings within a constrained vertical space. (*After Katz,*[8,9] *courtesy of Prof. J. Katz*)

TRAILING EDGE LIPS, NOLDERS OR ANGLE IRONS

One feature seen on some car wing designs is a strip of material added to the trailing edge of the wing upper surface, and bent up, usually at right angles. This device acts like a simple flap and increases the camber at the trailing edge, thereby increasing the lift coefficient. The arrangement provides a simple method of adjusting the down force.

TRAILING VORTEX (INDUCED) DRAG

Thus far, only the influence of the wing section geometry and the relationship between lift and profile drag have been described. The overall drag on a wing is however dominated by trailing vortex drag. In Chapter 2 an explanation was given of how wings and other lifting shapes produce trailing vortices which modify the air flow (see Fig. 2.10). The lift vector is tilted backwards to produce a drag component known as induced or trailing vortex drag. Trailing vortex drag is produced regardless of whether the lift is upwards or downwards. On damp days, the low pressure in the core of these vortices, which trail from the wing tips, causes the water vapour to condense, and the vortices can be clearly seen, as illustrated in Fig. 6.25. The process is described in more detail by Barnard and Philpott.[2]

Most domestic cars produce upward lift, and the direction of rotation of the trailing vortices is such that water spray in the wake is thrown out sideways as illustrated in Fig. 6.26(a). Racing car wings on the other hand produce downward lift, and the vortices rotate in the opposite sense. Water spray is therefore pushed

Fig. 6.25 On wet days, trailing vortices can often be seen in the cloud of condensed water vapour at the wing tips.
(*Photograph courtesy of LAT Photographic*)

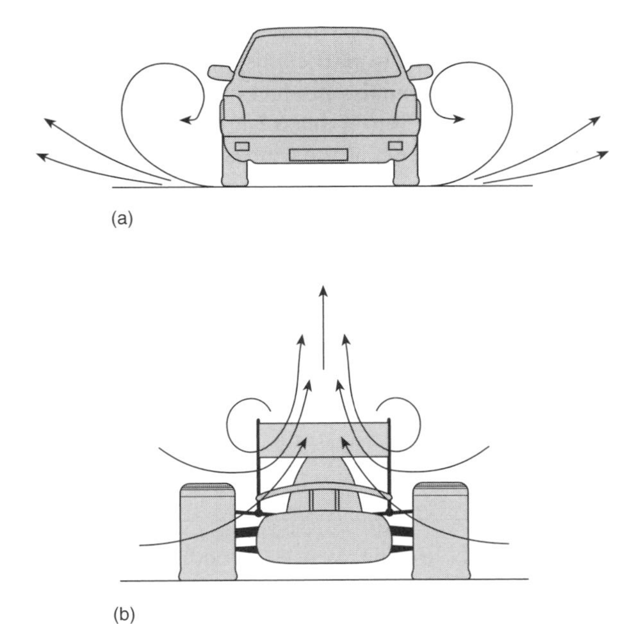

Fig. 6.26 Water spray patterns
Behind: (a) a domestic car generating positive lift; (b) a racing car generating down force.

upwards as illustrated in Fig. 6.26(b). The vertical plume of water that follows racing cars on rainy days can be quite spectacular.

The upwash (upward component of velocity) produced by these strong trailing vortices means that the approaching air is pulled upwards towards the wing, reducing its effective angle of attack. On open-wheeled cars, the flow approaching the wing is also subjected to an upwash produced by the rotating wheels (see Fig. 2.15), which further reduces the effective angle of attack. The reduction can be compensated for by increasing the incidence of the aerofoil, but this results in a further increase in drag.

WING PLANFORM AND CONFIGURATION

Trailing vortex drag is an unavoidable consequence of generating lift or down force using an isolated wing, but it can be minimized. The magnitude of the vortex drag must clearly be dependent on the strength of the trailing vortices, which in turn is related to the amount of lift being produced. The drag also depends on the wing aspect ratio: the ratio of the span to the chord (as defined in Fig. 6.15). The relationship is given by

$$C_D \text{ (trailing vortex)} = kC_L{}^2/\pi AR$$

where AR is the aspect ratio, and k is a constant which depends mainly on the wing planform geometry.

Racing car wings tend to produce very large C_D values because the C_L is high and the aspect ratio AR is low due to restrictions on width imposed by the racing regulations. A rectangular planform also produces a high k factor, and tapered wings are better in this respect. From Table 6.2 it can be seen that for a typical Formula 1 car, the drag contribution of the wing is large despite the modest wing size.

One advantage of low aspect ratio wings is that the flow is dominated by the trailing vortices, which induce high velocities over the rear portion of the aerofoil, thereby helping to maintain attached flow. The effect is similar to that described in Chapter 4, where the trailing vortices were shown to help maintain attached flow over the rear screen on hatchbacks. Keeping the flow attached enables the aerofoil to be set to high angles of attack, thereby generating high lift coefficients, albeit at the cost of increased drag.

END-PLATES

To reduce the influence of the trailing vortices on both lift and drag, it is normal to fit end-plates both on the front and on the rear wings, as shown in Figs. 6.21 and 6.22. The end-plates partially prevent the spillage round the tips and thus delay the development of strongly concentrated trailing vortices. The trailing vortices still form downstream, but near the wing itself the plates have the effect of distributing the trailing vorticity over a larger area. The result is that the upwash produced by the trailing vortices is less strong in the vicinity of the wing, and there is consequently less trailing vortex drag. If the plates are very large, the wing flow is almost the same as that for an infinitely long wing or a two-dimensional section, and there will be no trailing vortex drag or other upwash effects.

Although end-plates help, they by no means produce a cure for trailing vortex drag. They do have other advantages, however, in that they help to reduce the influence of the upflow from the wheels. They also provide a convenient and robust method of mounting the wings.

On open-wheel car designs, the end-plates of the forward wings were at one time curved and extended rearwards to separate the wing wake from the flow generated by the front wheels, but this was later banned on Formula 1 cars. Similarly, until restrictions were imposed, the front wing end-plates were extended down to the ground to provide side-skirts, which reduced inflow from the sides, and thereby improved the suction.

THE INFLUENCE OF GROUND AND BODY PROXIMITY ON WING SECTION SHAPE

The front wing is usually now mounted very close to the ground, and as with underbody down force, the proximity of the ground reduces the trailing vortex drag, and increases the down force. Figure 6.27 shows how ground proximity strongly influences the pressure distribution and overall lift coefficient. This figure shows clearly why it is important to take account of ground proximity when designing aerofoils for cars. If the pressure distribution in free-stream conditions in Fig. 6.27 were designed to just avoid separation in the adverse region, then the aerofoil would undoubtedly stall when close to the ground, because, as may be seen, the adverse pressure gradient becomes more severe. Katz[8] shows a similar effect for the case of a flapped aerofoil. Aerofoils for low-mounted front wings therefore have to be designed to give the required form of pressure distribution when in close proximity to the ground. The shape will be different from that of a section designed to operate with the same pressure distribution in free air.

In the early 1990s it became popular to mount the front wing under a relatively high nose fairing (Fig. 6.28). With this arrangement, the wing-in-ground effect works across the whole span, and the whole wing is exposed to the uninterrupted air flow. The high-mounted nose also permits a clean entry to the underside suction surface.

Because of restrictions on the mounting height, the rear wing is close to the bodywork and near the wake region at the back of the car. Both of these factors will affect the pressure distribution, as will the influence of the wheel-induced flow in open-wheeled cars. Figure 6.29 shows the effect of mounting a rear wing just above and behind the flat rear body panel of a sports car. It can be seen that the influence

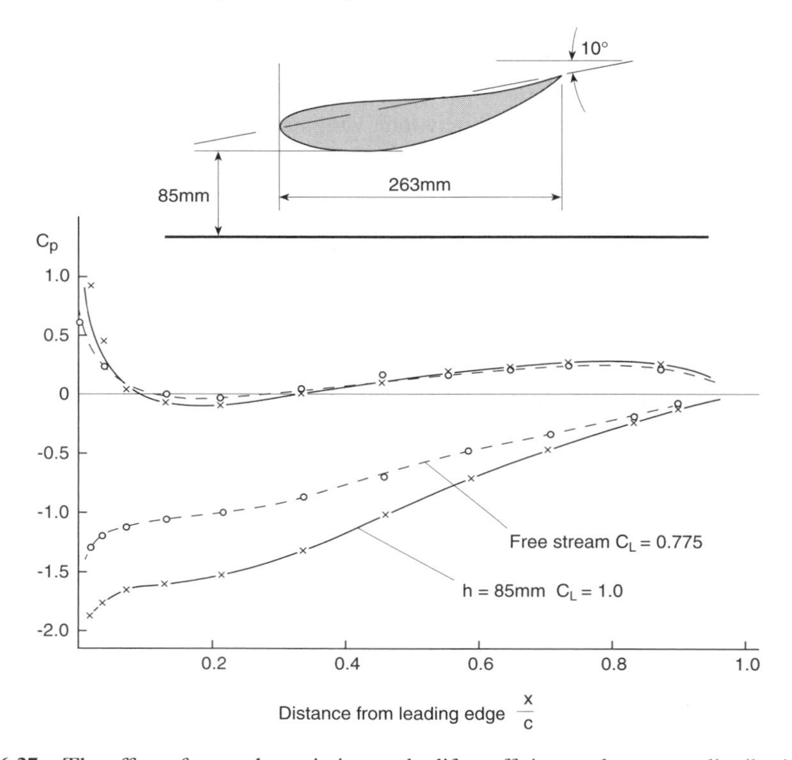

Fig. 6.27 The effect of ground proximity on the lift coefficient and pressure distribution around an aerofoil.
(*From data obtained in the University of Hertfordshire open-jet tunnel by Savill, Evans and Fisher*)

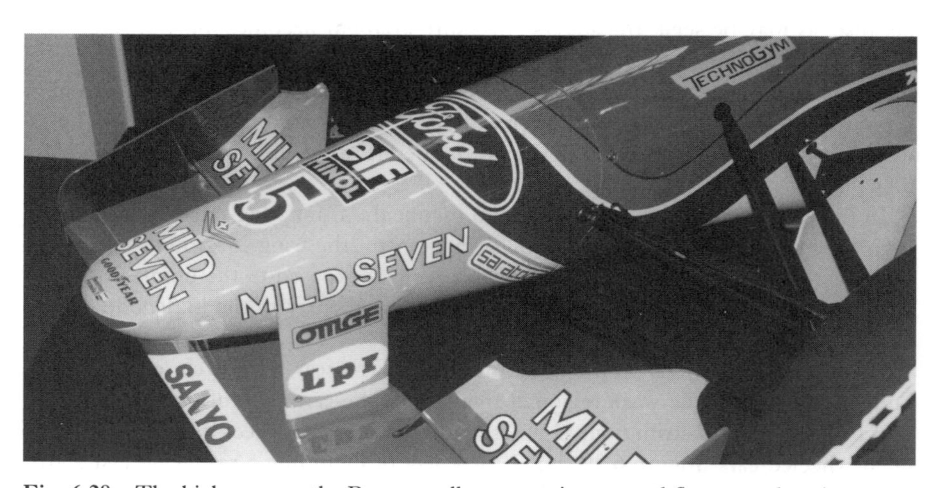

Fig. 6.28 The high nose on the Benetton allows an uninterrupted flow over the wing.

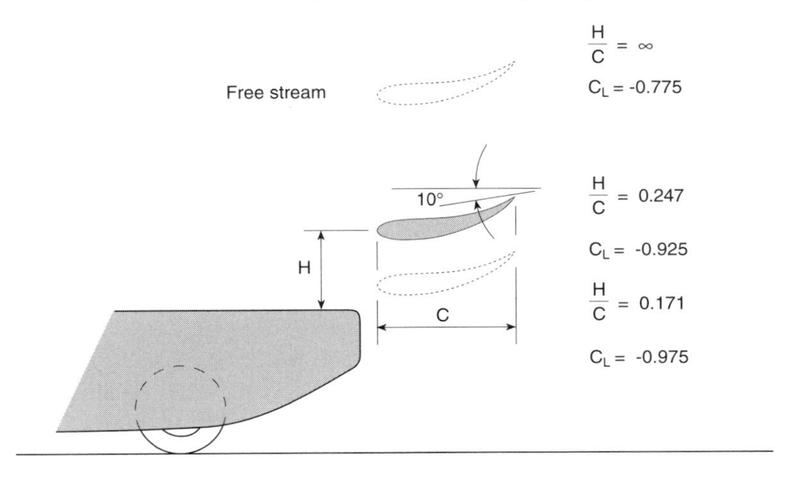

Fig. 6.29 The effect of interference from the rear of the body of a sports racing car on the overall lift coefficient.
(*From data obtained in the University of Hertfordshire open-jet tunnel*)

of the wing/body interference on the overall down-lift is unexpectedly beneficial. A similar effect was shown by Katz[8] for the case of a wing mounted just above the flat rear. It might be thought that mounting the wing close to the bodywork would always reduce the down force, but in the case of vehicles with strong ground effect, the low-mounted wing can help to reduce the underbody pressure, and hence improve the overall down force. Once again, interference effects must be considered when designing the aerofoil.

Computational fluid dynamics (CFD) techniques enable aerofoils to be designed to produce the desired pressure distribution in the realistic conditions of ground and bodywork proximity. However, despite the effectiveness of CFD methods, a great deal of wind-tunnel testing is still necessary in order to develop and verify the computed design. As described in Chapter 12, CFD results usually need to be tuned by using the feedback from experimental tests. Further information on racing car aerofoils and the design principles employed is given by Benzing,[3] Smith[13] and Katz[8].

UNDERBODY DESIGN FOR GROUND EFFECT

Current racing regulations for the major classes require that a large proportion of the vehicle's underbody should be flat. Figure 6.30 shows a typical flat-bottomed configuration. The underside, together with the ground, forms a channel through which the air passes. The mass of air flowing per second must the same all the way along the channel, unless there is some leakage in or out through the sides. (The

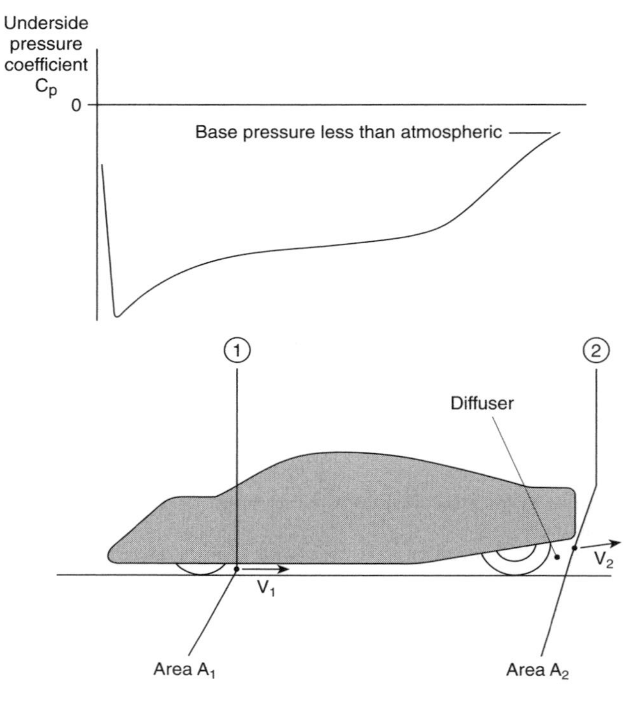

Fig. 6.30 The underbody flow channel, and a typical pressure distribution.

side-plates on the early designs were intended to prevent such an air leakage.) This continuity of air mass flow is expressed by the equation

$$\rho A_1 V_1 = \rho A_2 V_2$$

The terms are defined on the diagram Fig. 6.30. At normal racing speeds the compressibility of air is negligible, so the density remains constant. The volume flow rate $A \times V$ must therefore also be the same everywhere, and the above expression may be rearranged as

$$V_1 = V_2 \times (A_2/A_1)$$

From this it can be seen that under the flat section, where the cross-sectional area is small, the velocity will be higher than at the outlet from the diffuser, where the area is large. Bernoulli's equation states that where the speed is higher, the pressure must be lower, so the pressure at (1) must be lower than the pressure at outlet (2), which will be at roughly atmospheric pressure. The pressure under the vehicle is thus controlled by the ratio of the area under the body to the diffuser outlet area (as long as the flow remains streamlined and attached).

The length of the diffuser is limited by regulations, and its angle of divergence is restricted by the fact that if the angle is too great the flow will separate due to the adverse pressure gradient (see Fig. 4.34). These factors impose a limit on the outlet area from the diffuser, and therefore to obtain a large area ratio, and hence a large pressure ratio, the area under the floor pan must be kept as small as possible. Making this clearance small also helps to reduce the inflow of air from the sides, which would weaken the suction. Until restricted by regulations, the ground clearance on racing cars was at one time very small indeed, as may be seen in Fig. 6.8.

Because of the importance of the diffuser, and the restrictions imposed on its overall dimensions, a great deal of effort goes into designing effective diffusers, and fitting them around components. A variety of flow guide vanes and complex geometries may be seen on racing cars of all types. Figure 6.31 shows a good example.

A major problem with aerodynamic down force is that it increases with the square of the speed, so unless extremely stiff suspension springs are used, the ground clearance will vary with speed. At one time, active suspension systems were used in order to maintain the optimum small clearance, but these were subsequently banned.

It might be thought that it was the ratio of the front inlet height to ground clearance that controlled the pressure under the floor pan, but this is not so. Figure 6.32 shows three different inlet configurations. The pressure in the narrow channel will be roughly the same in all three cases. Only the flow and pressure distribution around that inlet region are modified. The inlet shape is however important in controlling the down force distribution. The high inlet in Fig. 6.32(a) produces an area of relatively high pressure at the front of the under-surface, and therefore pushes the centre of pressure of the down force rearwards. Experimental data for various front forms were obtained by Flegl.[5] The upturned nose gave a positive front wheel lift coefficient (C_L based on the front wheel lift) of 0.198, which was reduced to a negative (down force) value of -0.094 for the drooped nose.

Fig. 6.31 The exit from the underbody venturi can be seen on this Williams FW16 (1994). Note the guide vanes.

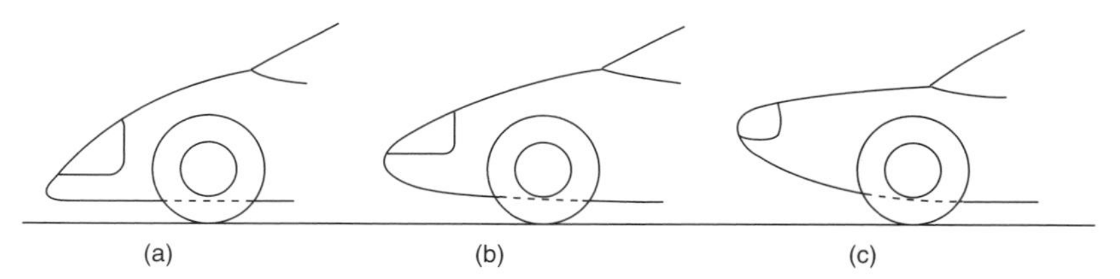

Fig. 6.32 Underbody channel inlet shapes
There is no advantage in trying to funnel the flow into the underbody area by using an upturned nose; all that happens is that there is a higher pressure under the nose, and the front wheel down force is reduced.

On open-wheeled racing cars it has become customary to extend the flat floor panel forward under a raised nose to form a splitter plate, as may be seen in Fig. 6.6, so there is no shaping of the underbody channel inlet.

On racing sports cars, it is difficult to produce enough down force on the front for a proper balance between front and rear adhesion, and the flat underside may be carried right up to the front of the body shell or even beyond, again often with no curvature in the inlet section (see Fig. 6.8). The diffuser sections are usually formed each side of the central engine/gearbox fairing. A small transverse groove near the front is sometimes used as a means of controlling the position of the centre of pressure.

Figure 6.30 shows a typical underbody pressure distribution. This figure helps to explain the important interaction between the rear wing and the underbody pressure. The wing tends to lower the pressure in the wake behind the vehicle where the venturi exit is located, and it is this pressure combined with the diffuser geometry that controls the pressure on the whole of the vehicle under-surface.

LOW DRAG – THE ADVANTAGE OF GROUND EFFECT OVER WING DOWN FORCE

Unlike wing-generated negative lift, down force generated by underbody suction does not normally carry a large drag penalty. Trailing vortex drag is produced as a consequence of the downwash or upwash velocities induced by the trailing vortices. If the lifting surface is close to the road, the presence of the road inhibits the formation of the vortices, as described in Chapter 2. The rubbing side-skirts that were at one time permitted were particularly effective in suppressing trailing vortex formation because they prevented the inflow from the sides, which is the primary origin of the associated trailing vortices.

The fact that underbody-generated down force does not carry a significant induced drag penalty is obviously a great advantage in comparison to wing-generated down force. According to Faul,[4] for ground-effect cars, down force/drag ratios of 300:1 can be obtained. This may be compared to the lift/drag ratios of around 40:1 which are realized on the best high-performance gliders. There will be some extra drag generated by ground effect, but this is largely due to viscous friction and pressure drag.

LIFT DUE TO EXPOSED WHEELS

Exposed wheels generate upward lift, which detracts from the down force generated by the wings and underbody. Stapleford and Carr[14] give a measured lift coefficient for four rotating wide racing-car wheels of around 0.6 (based on the frontal projected area of the wheels). About two-thirds of this lift is due to the front wheels, since the rear wheels are partly blanketed by the front pair. This positive (upward) lift can reduce the down force by around 11 per cent for a typical Formula 1 car set for a medium fast circuit. Table 6.2 earlier in this chapter gives some idea of the lift and down force balance due to the various components.

Enclosing the wheels within the bodywork as in a sports racing car reduces the wheel lift considerably. Clearly, if the bodywork fully covers the wheel so that it is not exposed to the air flow at all, no lift will be generated. Some care has to be taken with the form of the enclosure, however, since if the wheel is located within a large wheel-shaped bulge, as on some record-breaking cars (Fig. 4.11), the shape will correspond roughly to that of a non-rotating wheel, and Stapleford and Carr[14] indicate that the lift will be similar to that from a rotating wheel. The effects of partially enclosing a wheel have been studied by Scibor-Rylski.[12]

ROLLING RESISTANCE INCREASE DUE TO DOWN FORCE

As described in Chapter 3, rolling resistance is almost linearly related to the load or down force carried by each tyre. Aerodynamic down force will therefore increase the rolling resistance, and since aerodynamic forces increase as the square of the speed, so will the increment in rolling resistance. Accurate estimates of this effect are difficult because the rolling resistance coefficient varies in a rather complicated way with speed and with tyre temperature; however, for simplicity, taking the rolling resistance coefficient k_r as constant, the rolling and aerodynamic contributions to total resistance force F can be calculated from

$$F = \frac{1}{2}\rho V^2 A C_D + k_r(W - \frac{1}{2}\rho V^2 A C_L)$$

Putting in the data for a typical Formula 1 car from Table 6.2, and taking a weight of 6000 N (1350 lbf), it is found that at around 160 km/h (100 mph), the added rolling resistance due to aerodynamic down force contributes only about 2.5 per cent of the total resistance. In car racing, however, even such small contributions are important.

SLIPSTREAMING OR DRAFTING

When one vehicle follows another very closely, there is a mutual benefit in terms of reduced drag. In general, the following vehicle will derive the greater benefit, as may be seen from Fig. 6.33. The phenomenon may be explained by considering what happens when two block shapes are travelling in close convoy. In the limit, as they get closer, the pressure on the rear face of the leading vehicle will approach that on the front face of the following vehicle, so there will be a reduction in base drag (drag due to low pressure on the rear face) on the leader, and in front-surface pressure drag on the follower. Racing cars often make use of this in overtaking manoeuvres, the overtaking car taking advantage of the drag reduction that occurs as it comes close behind the one in front. Unfortunately, the overtaking vehicle also has to put up with strong slipstream effects, and usually a reduction in front down force. When aerodynamic devices were first introduced there were a number of catastrophic accidents which were attributed to aerodynamic effects. Howell[7] describes how several cars simply took off and flew, or flipped over backwards. Can-Am cars seemed to be particularly prone to this kind of incident, and Howell demonstrated in wind-tunnel studies that in close proximity to a leading vehicle, the follower could indeed generate sufficient front-end lift to flip over.

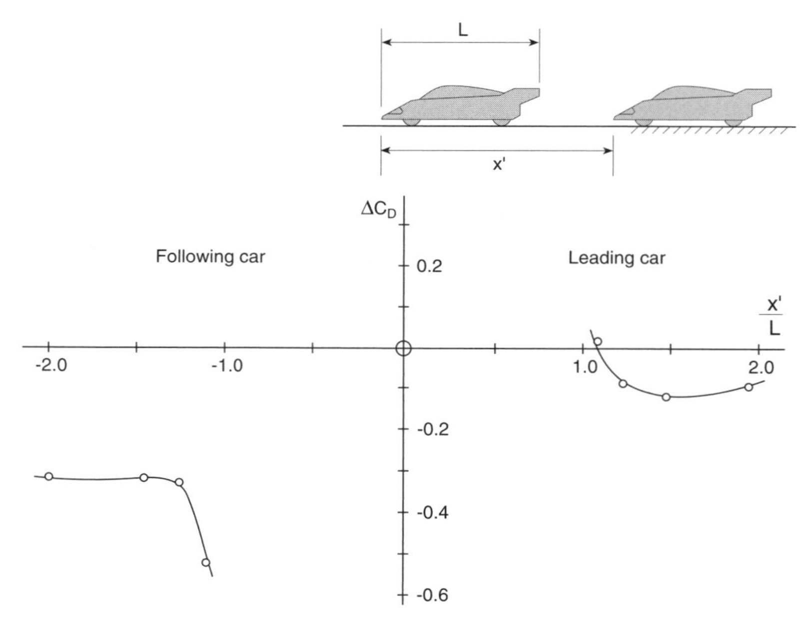

Fig. 6.33 Mutual drag reduction due to slipstreaming. The follower has the greater advantage.
(*After Howell,*[7] *courtesy of J. Howell, Rover Group Ltd*)

DOWN FORCE AND SAFETY

The increased cornering speeds which the use of aerodynamic down force allows increase the potential for disaster; spinning off a corner at 250 km/h (150 mph) is clearly much more dangerous than spinning off at 130 km/h (80 mph). Underbody down force is particularly dangerous in this respect, since once ground proximity is lost, most of the down force will disappear, and the vehicle will leave the track at very high speed.

Wing negative lift is less dangerous than underbody suction since it will usually continue to operate as long as the vehicle maintains a reasonably level attitude, and it is much less sensitive to small changes in vehicle attitude or height. Many changes in racing regulations have been aimed at reducing the down force contribution from underbody suction.

It should be noted that for road-going cars there are no restrictions on the use of movable aerodynamic devices, and it is possible to employ wings which increase their incidence when high down force is required: a technique used on early winged racing cars. Fortunately, on road-going cars, operational considerations dictate that the ground clearance should not be too small, and the potential for underbody suction and consequent disasters is limited.

RECORD-BREAKING VEHICLES

Record breaking used to be a relatively simple matter of producing as much power as possible, marrying this to a highly streamlined vehicle of low frontal area (Fig. 4.11), and then finding a very long area of flat ground. Nowadays, with turbojets and rockets, there is no shortage of power or at least thrust, and the main problems are with stability, and with having to cope with the change to supersonic flow. At the changeover from subsonic to supersonic flow, large alterations to the pressure distribution and the general flow field occur. At supersonic speeds, shock waves reflect from the ground and produce unstable discontinuities in the pressure distribution around the vehicle. Record-breaking vehicles have already exceeded supersonic speeds, demonstrating that the problems are not insurmountable. The problem of finding a suitable proving ground, however, remains. Acceleration to 1320 km/h (820 mph), the target for current attempts, takes about 27 s at an average of around 1.5*g* acceleration, during which time the vehicle will cover 5.5 km (approx. 3.4 mi). After the measured mile (1.6 km), during which a speed in excess of 1370 km/h (850 mph) is targeted, another 6.5 km (4 mi) is required for deceleration, making a total of around 13.6 km (8.5 mi) of totally flat terrain.

FURTHER ASPECTS OF RACING CAR AERODYNAMICS

It is impossible in a book of this size to describe all the aerodynamically driven design features found on racing cars, and many are now of purely historical interest. Some important aspects are covered in other chapters, notably cooling systems in Chapter 7. Until now, little mention has been made of popular racing car types such as competition modified saloon cars. These employ a mixture of pure racing car aerodynamic features, coupled with extensions of the drag- and lift-reduction devices such as air dams and spoilers that are found on production cars. Air dams may extend lower than on normal production vehicles, as the benefits of added down force more than compensate for any increase in drag. Other racing car classes follow the general aerodynamic principles developed on Formula 1, Indy Car, and Group C sports cars.

Nowadays, the level of scientific knowledge employed by the more successful racing teams is such that it is no longer sufficient to use intuition alone to make aerodynamic improvements. Any good idea has to be supported by computational or wind-tunnel experiments.

REFERENCES

1. Abbott I. A., and von Doenhoff A. E., *Theory of Wing Sections*, Dover Publications, New York, 1949.
2. Barnard R. H. and Philpott D. R., *Aircraft Flight* 2nd edn, Longman, Harlow, 1994.
3. Benzing E., *Ali/Wings* (in Italian with accompanying English translation), Automobilia, Milan, Italy, 1991.
4. Faul R., Ein Rennwagen steht im Windkanal, *Automobil Revue*, No. 41, Berne Oct. 2, 1980, pp. 37–9.
5. Flegl H., Die aerodynamische Gestaltung von Sportwagen, *Christophorus*, No. 98, 1969, pp. 18–19.
6. Gerhardt H. J., Kramer C. and Ammerschlager T., Aerodynamic optimization of a group 5 racing car, 4th Colloquium on Industrial Aerodynamics, Aachen 1980.
7. Howell J., Catastrophic lift forces on racing cars, 4th Colloquium on Industrial Aerodynamics, Aachen 1980.
8. Katz J., Considerations pertinent to race-car wing design, *Proc. Vehicle Aerodynamics Conference*, Loughborough University 18–19 July 1994, RAeS, pp. 23.1–7.
9. Katz J., *Race Car Aerodynamics*, Robert Bentley, Cambridge Mass., USA, 1995.
10. Liebeck R. H., Design of subsonic airfoils for high lift, *Journal of Aircraft*, Vol. 15, No. 9, 1978, pp. 547–61.
11. McGhee R. J. and Beasley W. D., Low speed aerodynamic characteristics of a 17 per cent thick airfoil designed for general aviation applications, NASA TN D-7428, Dec 1973.
12. Scibor-Rylski A. J., *Road Vehicle Aerodynamics*, Pentech Press, London, 1975.
13. Smith C., *Tune to Win*, Aero Publishers, Fallbrook Calif., USA, 1978.
14. Stapleford W. R. and Carr G. W., Aerodynamic characteristics of exposed rotating wheels, MIRA Rep. No. 1970/2.

CHAPTER 7

INTERNAL AIR FLOWS - ENGINE AND TRANSMISSION COOLING

A complete coverage of the design of vehicle engine cooling systems would alone be enough to fill a large book. This chapter is therefore restricted to an outline of the basic principles and requirements, and to a description of how the internal and external flows interact.

Apart from the ventilation of the passenger compartment, which will be dealt with in the next chapter, there are several air flows that are required in a road vehicle. In rough order of significance they are:

- engine water cooling
- engine oil cooling
- cooling of the transmission oil (particularly on automatics)
- cooling of the brakes
- intercooling of the turbocharger when fitted
- air inflow to the engine

THE ENGINE COOLING SYSTEM

For many years, the only significant perceived requirement of the engine cooling system was that it should be effective, but more recently, it has been realized that the cooling system can contribute appreciably to the overall drag of the vehicle and also to engine power losses associated with driving the cooling fan. According to White[13] as much as 4.5 kW (6 bhp) of engine power could be absorbed by the cooling fan of a medium-sized car at 112 km/h (70 mph) in older designs. The drag and power reduction factors are particularly important in the design of racing cars, and improvements pioneered in the racing field have fed back into domestic vehicle design. Good cooling system design can in addition contribute significantly to reductions in noise and weight.

Emmelmann[1] and Hucho[5] measured the drag contribution from the cooling system for a large number of domestic cars of the 1970s to 1980s era. The average increment of C_D was around 0.04. This would be enough nowadays to make the difference between a poor vehicle with an overall C_D of 0.34, and a typical one with a coefficient of 0.3. A similar order of cooling system drag was reported by Stapleford,[12] who found that the values were quite sensitive to yaw angle, the C_D increment dropping from 0.036 at zero yaw to 0.003 at 30°. No doubt the effectiveness of the cooling system would decrease accordingly.

The normal method of determining the cooling system drag is to measure the overall drag of the car in its normal condition, then subtract the drag measured with the radiator blanked off. Garrone and Masoero[3] took the unusual step of measuring the cooling system drag directly with a strain-gauge force balance system. They found that the direct measurements did not tie up with the changes in overall vehicle drag. This is because the cooling system affects the flow around the whole vehicle, so some of the drag increase appears in the form of changes elsewhere on the vehicle body. This finding emphasizes the fact that in road vehicle aerodynamics, it is unwise to try to totally separate the effects of different features.

THE MECHANISM OF DRAG PRODUCTION BY THE COOLING SYSTEM

Before going any further, it is necessary to explain how the cooling system can increase drag. At first sight, it might be imagined that making a hole in the front of the vehicle and allowing some air to pass through would reduce the drag. Figure 7.1 helps to show why it does not. The simple streamlined shape shown in this figure will have quite a low pressure drag, with much of the drag coming from surface friction. By allowing air to flow through it, the wetted surface area is

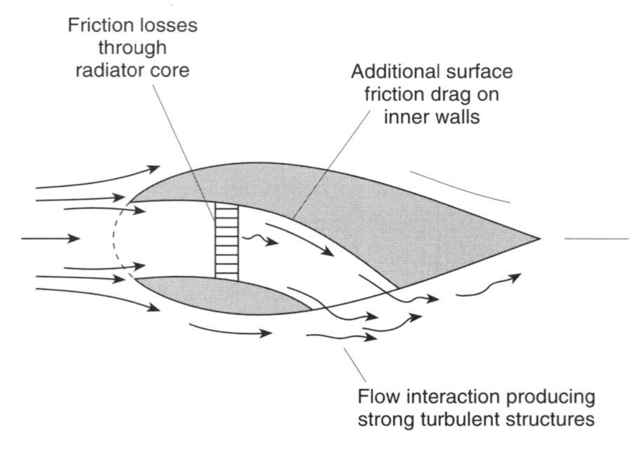

Fig. 7.1 Making a hole through a vehicle increases the drag due to the extra internal friction, and by causing interference with the external flow.

enlarged, increasing the surface friction drag. More importantly, drag is produced by flow separation and turbulence in the outlet area. If a radiator is inserted in the duct, there will be further friction losses, caused by the flow over the cooling fins. The whole purpose of a radiator is to provide a large surface area in order to conduct heat into the air. The overall surface area of the tubes and fins of the radiator can be as much as 100 times as great as its frontal area, so there will be considerable surface friction drag even though the flow speed through the radiator is lower than the free-stream speed.

Figure 7.2 shows a traditional domestic car engine cooling system. The air had to pass around many obstructions and bluff shapes as well as through the radiator. The decorative grille at the front often provided a quite unnecessary pressure loss, although it is not difficult to design a visually attractive inlet which provides a streamlined means of guiding the flow into the radiator. On traditional designs there was often a noticeable loss of stagnation pressure through the inlet grille in addition to the large loss through the radiator core. Paish and Stapleford[7] found that grilles on vehicles of the late 1960s were reducing the cooling air flow by between 13 and 59 per cent.

A loss of stagnation pressure is important, because it represents a loss of *available* energy. Energy cannot really be lost, but it can be converted into a useless form such as turbulent energy, from which recovery to a more useful form is virtually impossible. Note that changes in the static pressure are less important; as long as there is no loss in stagnation pressure, the static pressure simply reflects changes in the flow speed, as given by the Bernoulli relationship. The static pressure can be 'recovered' by slowing the flow down.

Another cooling system influence on drag is that the outlet flow can interfere with the external flow. This can result in strong turbulence with an attendant increase in drag, but conversely, it is also possible sometimes for the emerging air to help maintain attached flow, thereby reducing drag.

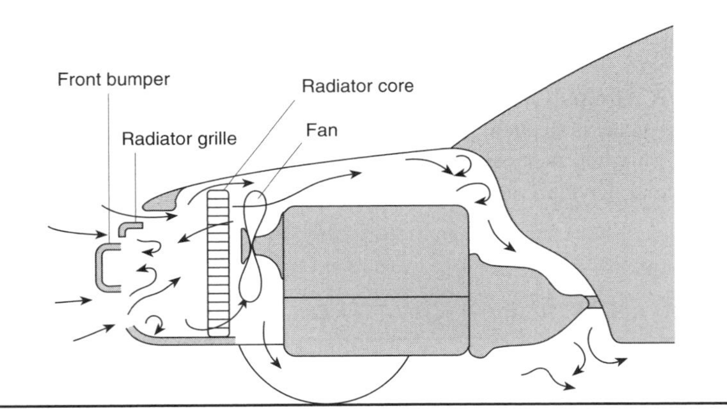

Fig. 7.2 A traditional domestic car radiator system of the unducted 'free-flying' type.

COOLING REQUIREMENTS

A spark-ignition (petrol) engine runs at a maximum efficiency in the range of about 30–40 per cent; in other words, only about one third of the energy available in the fuel is turned into useful work. If the efficiency is only around 33 per cent, it follows that the other 67 per cent of the available energy in the fuel must be dissipated as heat. Typically, up to about 30 per cent of the energy supplied by the fuel goes into the cooling water, the remaining fraction being lost with the hot exhaust gases and from the surface of the engine. This means that a typical 1.6 litre car engine producing 63 kW (85 bhp) of power has to dissipate an equal amount via the cooling system: equivalent to the output from 63 electric fire heating elements. On a racing car producing anything up to around 600 kW (800 bhp), the cooling requirements are more than an order of magnitude greater.

Domestic vehicles are rarely driven at maximum power for long periods, and cruising at 112 km/h (70 mph), a medium-sized family car will only require about one third of its maximum power output. The engine efficiency under part-load drops, however, so a considerable amount of heat still has to be dissipated in the cooling system. Diesel engines are more efficient than spark-ignition units, and a higher proportion of the waste heat goes into the exhaust gases, so the water cooling requirements are lower for a given output power.

It is generally considered unnecessary to design a cooling system that will cope with the most extreme requirements of, for example, maximum power at low road speed under tropical conditions; the design points usually chosen are (1) continuous operation at maximum speed and weight, and (2) climbing a continuous hill of 10 per cent gradient at 25 km/h (16 mph) at maximum loaded weight, towing a maximum permissible weight of trailer.

Most cooling systems are designed to provide sufficient cooling by ram air alone (i.e. without fan assistance) under normal cruising conditions, but at low speeds or when climbing hills etc., a fan is required. Track racing cars usually have no cooling fan, and rely entirely on ram pressure.

At one time, the cooling fan was coupled directly to the engine, an inefficient arrangement, because at high road speeds, where the fan speed is high, little or no fan assistance is required, whereas whilst stationary at tickover on a hot day, a great deal of fan assistance is needed. Nowadays most cars use an electric fan which is only switched on when necessary. The efficiency of the electric fan system is low, but under normal driving conditions it may rarely be needed.

DESIGNING AN EFFICIENT LOW-DRAG COOLING SYSTEM

In order to overcome the friction drag of the radiator, there must be a pressure difference between the upstream and downstream sides of the core. In the absence of compressibility effects, this pressure drop represents a drop in stagnation pressure that cannot be recovered. It is common to relate the pressure drop Δp to the dynamic pressure of the flow through the core ($\frac{1}{2}\rho V_c^2$) by

$$k_p = \Delta p / \tfrac{1}{2}\rho V_c^2$$

k_p is not quite a constant; at very low flow speeds, the boundary layer on the core elements is often partly laminar, and in this case k_p, which is directly related to the surface friction drag coefficient, would be proportional to 1/(Reynolds number), and hence to $1/V_c$, where V_c is the flow speed through the core. At higher speeds typical of a car cruising on the open road, the flow tends to become fully turbulent, and following classical boundary layer theory, k_p becomes proportional to roughly $1/V_c^{1/5}$. Hoerner[4] shows that above a core flow speed of 4 m/s the experimental data for k_p fits a curve of $1/V_c^{1/5}$ very well.

Although k_p is really a variable, the variation over the speed range of interest is not large; Hoerner's data show a variation of around 12 per cent between speeds of 8 and 16 m/s, and for practical convenience k_p is nearly always taken to be constant. k_p values for car radiators are in the region of 4–10, with 6 being typical for a moderate density core. The more cooling fins there are per square metre of core, the higher the loss coefficient will be.

UNDUCTED RADIATORS

Figure 7.3 shows the enclosing stream tube shape for an unducted or 'free-flying' cooling system, which corresponds roughly to the arrangement in older vehicles. The air flowing through the radiator starts with the free-stream conditions of pressure and speed and returns to the free-stream pressure downstream, but at a lower speed. The fact that the speed must be lower can be seen from Bernoulli's equation modified to take account of the stagnation pressure loss Δp that occurs in the radiator core.

$$p(\text{upstream}) + \tfrac{1}{2}\rho V^2(\text{upstream}) - \Delta p = p(\text{downstream}) + \tfrac{1}{2}\rho V^2(\text{downstream})$$

The expression above shows that if the upstream and downstream pressures are equal, then the speed at outlet must be lower than at inlet, and therefore the stream tube must be larger at outlet in order to pass the same quantity of fluid.

The pressure field causes the flow to slow down and therefore spread as it approaches the core. The relationship between the loss coefficient k_p and the slowing down of the flow at the radiator is given by Hoerner[4] as

$$\frac{\text{radiator speed}}{\text{vehicle speed}} = \frac{(V_c)}{(V)} = \frac{1}{1 + k_p/4} \qquad [7.1]$$

where V_c is the speed in the core, and V is the free-stream speed.

From the above, it will be found that with a loss coefficient k_p of 8, the velocity ratio is $\frac{1}{3}$, which means that if the intake aperture is placed well upstream, it would only need to be one third of the size of the radiator. Figure 7.3 shows the

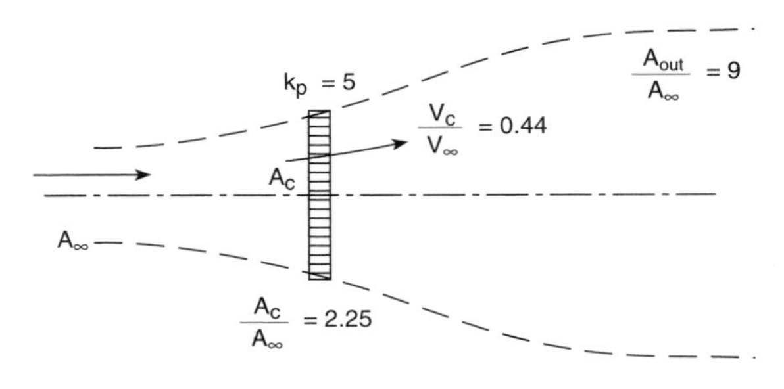

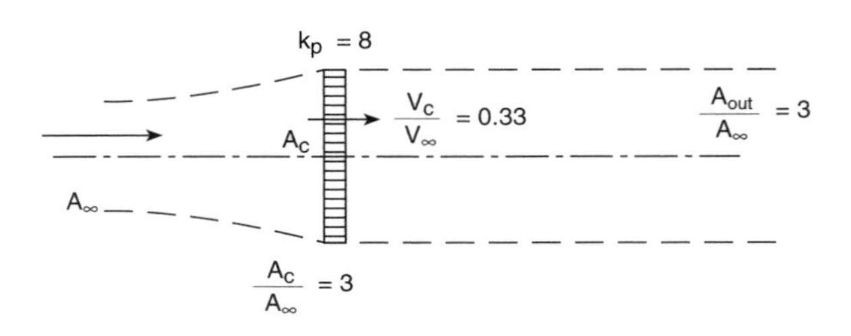

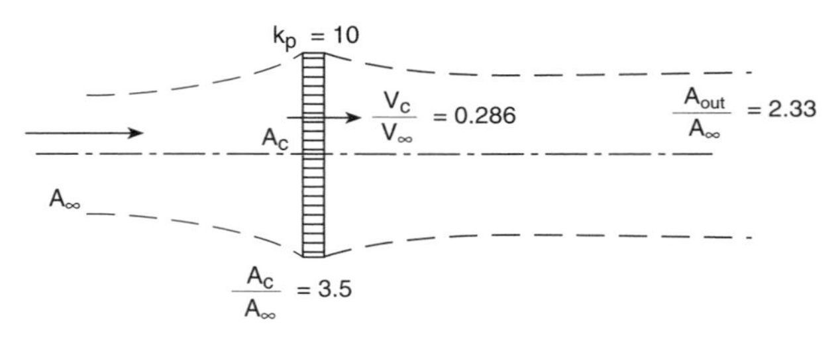

Fig. 7.3 The enclosing stream tube shape for an unducted or 'free-flying' cooling system, which corresponds roughly to the arrangement in pre-war cars. The air flowing through the radiator starts with the free-stream conditions of pressure and speed and ends up at nearly the free-stream pressure, but at a lower speed.

calculated surrounding stream tube dimensions for radiators with different values of k_p. It may be seen that with $k_p = 8$, the downstream area is the same as the core area, while with a higher loss coefficient, the stream tube narrows on exit.

DRAG ON THE RADIATOR CORE

The drag on the radiator core is mainly due to surface friction drag, which is proportional to the *local* dynamic pressure ($\frac{1}{2} \rho V_c^2$) and the surface friction drag coefficient, the latter being directly related to k_p.

A rather unexpected consequence is that the drag tends to *decrease* as the core becomes more densely packed with cooling fins. This is because of the resulting reduction in flow speed through the radiator V_c. Though surprising, this is consistent with the fact that when the radiator is fully blocked or blanked off, the overall vehicle drag decreases.

DRAG REDUCTION USING DUCTED RADIATOR SYSTEMS

A reduction in the speed through the core is possible if the radiator is enclosed in a duct, as shown in Fig. 7.4. Following Bernoulli's relationship, as the duct area increases, the flow speed decreases. The slowing down effect is useful because, as shown above, the pressure loss, which is responsible for the drag on the radiator core, is dependent on the flow speed in the core.

For a ducted radiator system, Hoerner[4] derives an expression for drag coefficient (based on the area of the core A_c, and the *free-stream* dynamic pressure $\frac{1}{2}\rho V^2$). Taking the simple case where the outlet pressure is equal to the free-stream pressure, his expression reduces to

$$C_D = 2(V_c/V)(1 - (1 - (V_c/V)^2 k_p)^{1/2}) \qquad [7.2]$$

This expression is plotted against velocity ratio core speed:free-stream speed (V_c/V) in Fig. 7.5 for three different values of the radiator loss coefficient k_p. The values corresponding to an unducted radiator are also indicated. It will be seen that a relatively small change in the velocity ratio makes a large reduction in the drag coefficient.

Using the expressions above, it is possible to obtain an idea of the order of drag reduction possible by using a ducted radiator. Starting with the case of an unducted

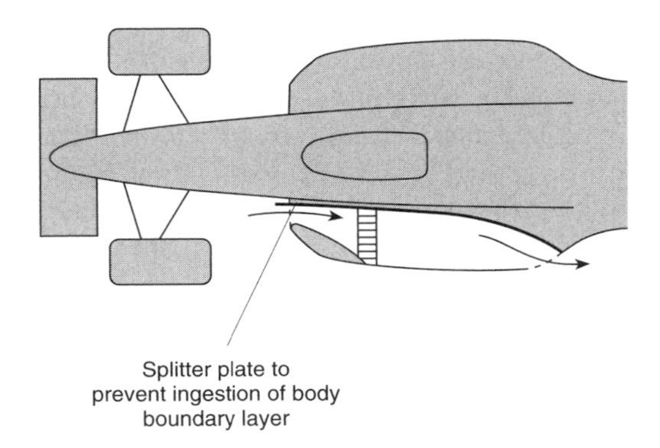

Fig. 7.4 A reduction in the speed at the core is possible if the radiator is enclosed in a duct, as on most current track racing cars.

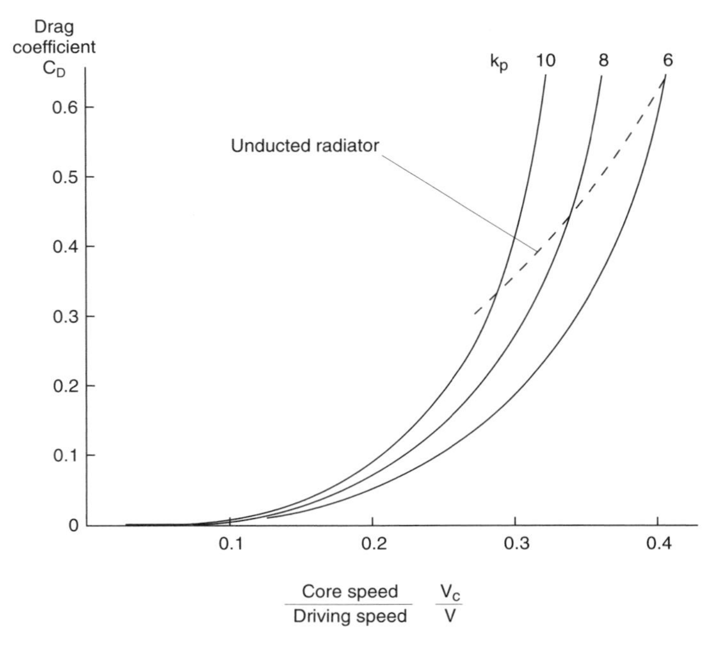

Fig. 7.5 Equation [7.2] plotted against velocity ratio (V_c/V) for three different values of the radiator loss coefficient k_p. The values corresponding to an unducted radiator are also indicated. It will be seen that a relatively small change in the velocity ratio makes a large reduction in the drag coefficient.

radiator with a $k_p = 6$, we can see from Fig. 7.5 that the drag coefficient is 0.64, and the velocity ratio V_c/V is 0.4. Reducing the velocity ratio to 0.35 lowers C_D to 0.34, a factor of 0.53.

The drag depends not just on the drag coefficient, but on the product of the drag coefficient and the core area ($C_D A_c$), however, and to maintain the original rate of cooling despite the lower air speed, it is necessary to increase the radiator area. The amount of area increase depends on whether this is achieved by a change in the height or a change in the width of the core. If both are increased in equal proportion, then the required area increase is roughly proportional to $(1/V_c)^{0.8}$. Despite the increase in area, there is still a net loss of drag, because the fall in C_D more than compensates for the increase in A. In the above example, the increase in radiator size required would be $(0.4/0.35)^{0.8} = 1.113$. The change in $C_D A_c$ would therefore be 1.113×0.53, which is 0.59.

Figure 7.6 shows the variation of drag with core speed:free-stream speed ratio if the radiator size is adjusted to maintain a fixed amount of cooling. The starting reference point is an unducted radiator with $k_p = 6$. It will be seen that using a duct to slow the air down by even a modest amount can produce a useful reduction in cooling system drag. The true reduction would be a little less, because, as shown previously, k_p is not really constant, but proportional to $1/V_c{}^{1/5}$.

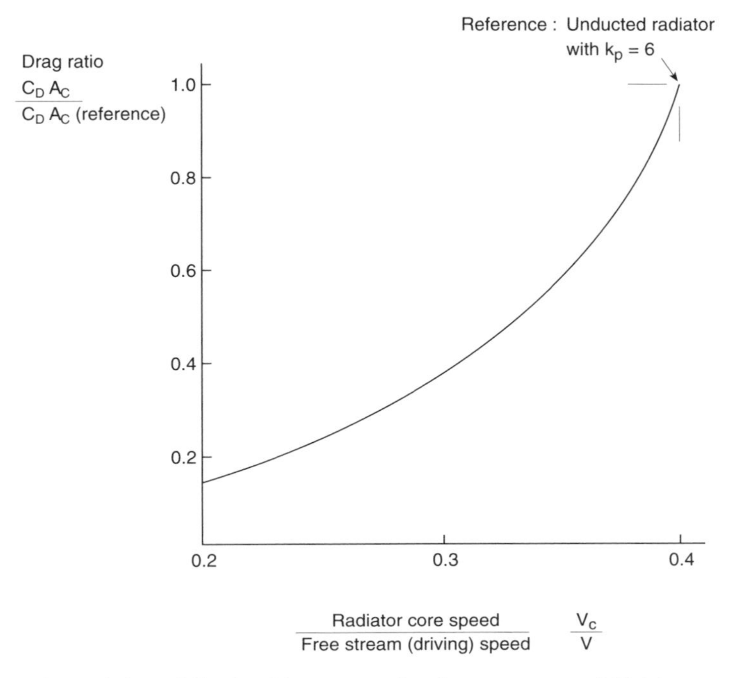

Fig. 7.6 The variation of $C_D A_c$ with core speed to free-stream speed (driving speed) ratio, for a constant rate of cooling.

It should be noted that in a ducted design, the overall drag cannot simply be calculated from the drag on the radiator core; the duct walls are not at atmospheric pressure, and can actually make a negative contribution to drag. In the above example, although the core area was used in the calculation, the drag is that of the overall ducted system.

The optimum duct arrangement depends on striking a balance between drag reduction and the size, cost and weight penalties of using a large radiator. Furthermore, with a high ratio between the free-stream and radiator flow speeds, it becomes very difficult to maintain attached flow in the intake diffuser, since the flow is going against an adverse pressure gradient. If separation occurs, a large amount of turbulence will be generated, and the flow will probably reduce to a free jet flowing though only part of the core (Fig. 7.7).

As with the flow through the underbody venturi on racing cars, the flow through the cooling duct is largely controlled by the exit pressure and outlet venturi shape. Once the outlet conditions, outlet duct shape and radiator area and loss coefficient have been fixed, changing the intake duct will not alter the flow, unless it starts to become choked, or the flow becomes separated from the walls; the stream tube simply adjusts, as shown in Fig. 7.8.

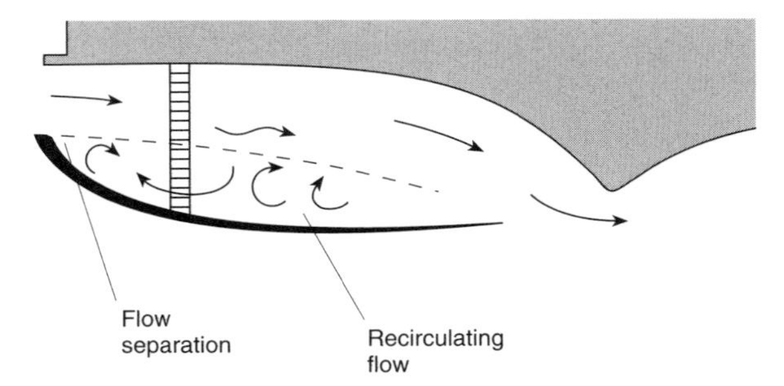

Fig. 7.7 With a large ratio between the free-stream and radiator velocities, it becomes difficult to maintain attached flow in the intake diffuser, since the flow is going against an adverse pressure gradient. If separation occurs, a large amount of turbulence will be generated, and the flow will probably reduce to a free jet flowing though only part of the core.

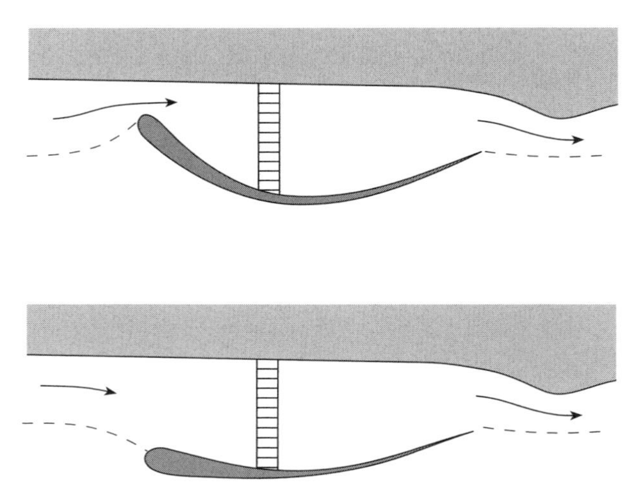

Fig. 7.8 Once the outlet conditions, outlet duct shape and radiator area and loss coefficient have been fixed, changing the intake duct will not alter the flow, unless it starts to become choked, or the flow becomes separated from the walls; the stream tube simply adjusts.

NEGATIVE DRAG

A great deal of thermal energy is transferred into the air stream from the cooling system, and it is possible to use the cooling system duct as a form of ramjet, to produce thrust or negative drag. This was used to some effect on high-performance piston-engined aircraft, but unfortunately, the efficiency of the ramjet's theoretical thermodynamic cycle, like that of the reciprocating engine, depends on the pressure ratio. If ram air compression alone is used; i.e. the pressure of the air is raised to near its stagnation value by nearly bringing it to rest, then for a car at a road speed of 112 km/h (70 mph) the theoretical maximum efficiency of the ideal ramjet thermodynamic cycle is only around 0.15 per cent. Thus, if the engine was producing say 30 kW (40 bhp), and an equal rate of energy was transferred to the cooling water, the maximum additional thrust obtained would be equivalent to less than 45 W (0.06 bhp): only enough power for a small light bulb. Nothing like even this trivial amount could be realized in practice, so for a domestic vehicle the effect is not worth bothering with. For a racing car at 320 km/h (200 mph), however, the maximum theoretical efficiency rises to around 1.5 per cent, so a marginal effect might be realizable, and in racing, every little counts.

DOMESTIC CAR ENGINE COOLING SYSTEM DESIGN

For a domestic car, the main requirements are that the air should flow cleanly through the system without encountering unnecessary obstacles, and that the flow speed should be reasonably uniform across the face of the core. This latter condition is not at all easy to achieve.

Even if a fully ducted system is not used, there are significant advantages in using a ducted intake section designed simply to prevent flow separation. There have been several studies of the effect of the inlet configuration. Renn and Gilhaus[8] show measurements of the drag increase due to the cooling system plotted against engine power for both ducted and unducted intake configurations. There is a significant amount of scatter in the data, which indicates that in the early 1980s there was a wide range of effectiveness in cooling system design. The ducted intake designs, however, showed a clear advantage, with an average $C_D A$ value of 0.21 m^2 compared with 0.465 m^2 for the unducted designs. Detailed investigations showed that by using a ducted intake, as opposed to a free-flying arrangement, the drag due to cooling could be halved, and the flow velocity increased by 25 per cent, with more uniformity across the face of the core. The reduction in drag was also shown to be accompanied by a useful decrease in front-end lift. Figure 7.9 shows just how small the inlet aperture needs to be on a well-designed ducted intake.

Renn and Gilhaus[8] and Oler *et al.*[6] also show the influence of the geometry and positioning of the intake on conventional domestic car designs. It is unfortunate that for legal reasons, the front bumper usually ends up being directly in line with the radiator core, and this necessitates having effectively two intakes, one above the bumper, and the other below, as may be seen in Fig. 7.9. The bumper in this case appears as an integral part of the bodywork. It is shown in the above two references that by experimental development of the intake geometry, it is possible to improve the velocity distribution and reduce the loss in stagnation pressure significantly.

With ducted designs, it should be possible to eliminate losses in stagnation pressure in the intake almost completely. All that is required is that the flow should not encounter any bluff obstructions such as decorative grilles, and that it should remain attached to the duct walls right up to the radiator. If this is achieved, then the only losses will be those due to surface friction, which will be small. Most older

Fig. 7.9 The production Ford Probe showing a neat, streamlined, and relatively small cooling air intake aperture. The positioning of the front bumper (integral in this case) often breaks the intake aperture into two parts.

intakes acted much like a sharp-edged orifice, with large losses in stagnation pressure due to the separated flows. In good recent designs, the intakes and baffles are much more streamlined. The cooling fan or fans are usually electrically driven, and shrouded to increase their effectiveness and to reduce noise. Paish and Stapleford[7] describe an experimental ducted fan and radiator system in a family saloon car.

RACING CAR WATER AND OIL COOLING SYSTEM DESIGN

Except when used in modified production car classes, racing car cooling systems are now almost always of the fully ducted type. Cooling the oil is as important as cooling the water, and large oil cooler radiators are required in addition to the water cooler. Formula 1 and Indy Car vehicles have adopted the technique of placing the radiators in pods each side of the cockpit, as may be seen in several of the photographs in Chapter 6.

A feature seen in many racing vehicles is the use of an inclined radiator as illustrated in Fig. 7.10. Inclining the radiator allows a larger core to be contained in a given size of duct. It also means that the flow speed through the core is reduced, since $V_c = V_a \cos\alpha$. The difficulty with this arrangement lies in trying to obtain an even distribution of flow speed across the face of the core. A useful study of inclined radiators is described by Rivers *et al.*[9] It was found that for inclinations of up to 30°, the drag increased, and the heat transfer deteriorated due to the uneven flow distribution and general disturbances. Above this angle, however, both aspects improved, and at 55° there was a drag reduction of 20 per cent and a heat transfer

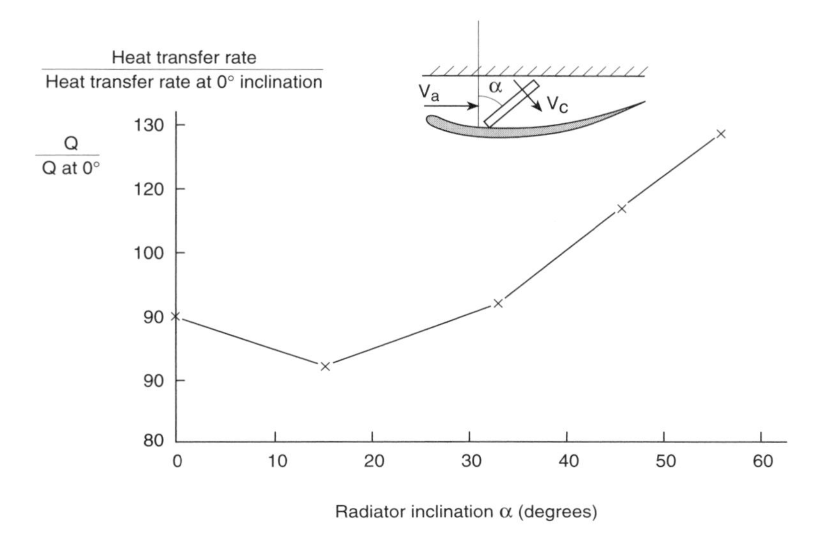

Fig. 7.10 The effect of inclining the radiator
Inclining the radiator allows a larger core to be contained in a given size of duct. It also means that the flow speed through the core is reduced since $V_c = V_a \cos\alpha$.
(*After Rivers et al.*[9])

increase of 30 per cent. There is no evidence of an upper limit to the angle of inclination, but the advantages are progressively offset by the increase in weight due to the use of a larger radiator.

As described earlier, it is the outlet pressure and area that really controls the flow through the duct, and ideally the outlet area should be varied with the speed, as was done on large piston-engined aircraft. Movable aerodynamic devices are not allowed on most classes of racing cars, but it is usual to include an adjustable flap which is fixed whilst under way after being set for the best compromise. On ordinary road vehicles, there is no reason why a variable outlet geometry should not be used, and the technique has been used on experimental low-drag vehicles such as the Ford Probe V.

INTAKE LOCATION

There is considerable scope for variety in the design of cooling systems, but except in some racing car classes, it is still common to place the radiator and intake in the front, and in this case, exhausting the air through a slot on the top of the bodywork in front of the windscreen seems to be the best configuration in terms of low interference drag and good down force. This arrangement is quite common on high-performance and competition vehicles.

Fig. 7.11 Air intake and outlet on top of the wheel fairings of a wind-tunnel model of a Group C sports racing car. The oil radiator is housed on one side, and the water radiator on the other.

The front intake has disadvantages in cars with a very low ground clearance, because the air temperature just above the road surface may be very high on hot days. On sports racing cars, the air is often taken in through an aperture in the side or top of the vehicle. Figure 7.11 shows the intake for a group C car model; the intake is on the top of the wheel fairing, and the outlet is not far downstream. This arrangement is not ideal, as the intake is in an area of rather thick boundary layer, but in racing car design it is necessary to balance many conflicting requirements.

THE NACA DUCT

When a side- rather than a forward-facing aperture is required as in Fig. 7.11, an aerodynamically good form is the NACA duct shown in Fig. 7.12. This type of intake was developed by NACA, the American National Advisory Committee on Aeronautics, forerunner of NASA. It relies on the fact that as long as the air can be taken from the free stream without energy losses, there is no disadvantage relative to a forward-facing aperture. The edges of the slot are sharp, and the resulting separated flow rolls up into a conical vortex along each side, which helps to draw the flow in. There are small losses associated with the fact that some low-energy air from the upstream surface boundary layer is ingested, and it follows that the duct should be mounted in a region of attached flow: preferably in a region near the front, where the boundary layer is thin. Apart from the need for sharp edges, a precise geometric shape is required, and the slope in profile should be less than 10°. The geometry is given in Smith[11]. Figure 6.8 in the previous chapter shows good examples of NACA ducts on the front of a Group C sports car. In this case

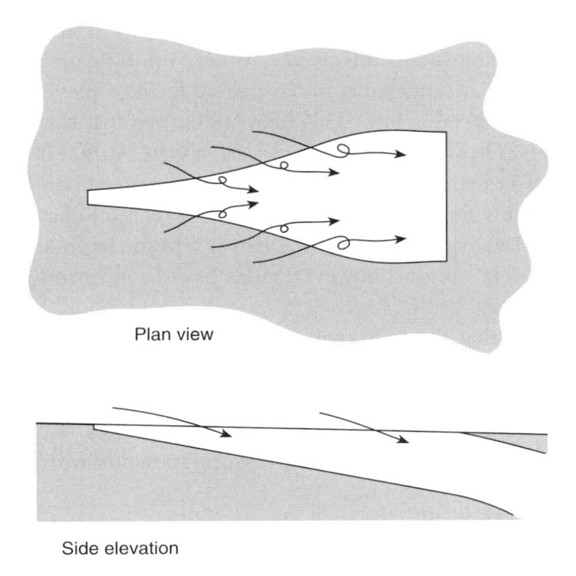

Fig. 7.12 The NACA duct
To work correctly, the geometry has to be precise, and the edges must be sharp so as to generate trailing vortices.

they are used for cooling the driver and the brakes. The water and oil radiator intakes are on top of the wheel fairing, as in Fig. 7.11.

One problem with the NACA duct is that it is rather large and long in relation to the amount of air that it can ingest. This leads designers to produce modified shapes, often with disappointing results. The intake in Fig. 7.11 would appear to be far from ideal, in that it is in a region of rather thick boundary layer, and it does not conform to the NACA duct dimensions. However, in real-life design, there are many conflicting constraints, and compromises are necessary. NACA ducts are commonly used for the smaller volume cooling flows such as for brake cooling on racing cars.

AIR BOXES AND RAM AIR INTAKES FOR ENGINES

When a vehicle is in motion, the engine intake pressure can be raised by an amount equal to the ram air pressure: the dynamic pressure of the undisturbed stream ($\frac{1}{2}\rho V^2$). This has a supercharging effect, and at very high speeds can make a noticeable difference to the engine output. For example, at 83 m/s (300 km/h or 188 mph) the ram air pressure is $\frac{1}{2}$ 1.226 83^2 = 4223 Pa (0.61 psi). Note, however, that at the common European legal maximum speed of 130 km/h (81 mph), the ram air pressure is only 799 Pa (0.11 psi): about as much help as a putting the emblem 'GT' on the rear.

To obtain the ram effect, the air has to be taken cleanly from the free stream without friction or losses due to turbulence. Where multiple air inlets are used, as on racing engines, it is advantageous to have some form of plenum chamber, a large volume where the air speed is low. This helps to ensure that the air is distributed evenly to all intakes. On some racing car classes, a NACA duct is commonly used, but as may be seen in many of the photographs in Chapter 6, on Formula 1 cars it was customary to use a forward-facing aperture and plenum chamber known as an 'air box' located just above and behind the driver's head. In an attempt to reduce engine power for safety reasons, however, rules have been introduced aimed at reducing the ram effect.

It should be noted that the only requirement for a good intake is that it and the associated duct should not cause energy losses, and it should be located in a region where the external air flow has not suffered a loss in stagnation pressure. The intake should therefore not be located in a wake or in a thick boundary layer. Care has to be taken to avoid strongly adverse pressure gradients within the duct, as these could lead to flow separation.

Consideration of Bernoulli's equation shows, perhaps surprisingly, that the full ram pressure can still be obtained even if the air is taken from an area of low *static* pressure. What matters is that the *stagnation* pressure should not be lower than in the free stream. It is hard to get people to believe that having a low static pressure in the intake does not alter the ram effect, as long as the stagnation pressure has not been lowered. An interim regulation aimed at reducing the ram effect on racing cars stipulated that the rear of the air box should be opened, thus lowering the static pressure. However, the only way to guarantee a reduction in ram air pressure is to insist that the intake should be located in a wake region.

OTHER AIR SUPPLY REQUIREMENTS

Apart from the cooling of the engine water and oil, automatic gearboxes need a heat exchanger, which is often located integral with the water radiator. High-performance and racing cars also require cooling of the gearbox, brakes, and turbocharger intercooler when fitted. Roussillon[10] gives the following air volumetric flow rate requirements for a Le Mans GT prototype category racing sports car at 360 km/h (225 mph!).

engine water radiator	6 m^3/s (210 ft^3/s)
engine oil radiator	3 m^3/s (100 ft^3/s)
gearbox oil radiator	1 m^3/s (35 ft^3/s)
turbocharger intercooler	0.8 m^3/s (30 ft^3/s)

No figures are given for the air supply to the brakes.

Flegl and Bez[2] give the engine cooling air flow requirement at full power as $1.39\,m^3/s \pm 0.4\,m^3/s$ per 100 kW ($49.0 \pm 14.1\,ft^3/s$ per 134 bhp) of installed engine power.

AIR FLOW IN THE ENGINE COMPARTMENT

In present domestic car designs, the air from the radiator passes around the engine block and provides a certain amount of cooling for accessories such as the alternator. Williams *et al.*[14] have studied the flow in an engine compartment using both model and full-scale vehicles in water tanks. The flows were found to be highly complex.

The experimental techniques required for such studies are by no means simple, and efforts are being made to model the flows using computational fluid dynamics (CFD). The problems of generating a computational grid to represent the complex and cramped geometry of the engine compartment adequately are considerable, and to this must be added the complication of the effects of heating. Nevertheless, the benefits of being able to use CFD are obvious, and developments are continuing in this area.

DESIGN FEATURES FOR WELL-DESIGNED COOLING SYSTEMS

Summarizing the above, the engine cooling system should preferably be of the fully ducted type, but in the absence of a fully ducted design, a well-shaped intake duct should at least be provided. Although the size of the inlet aperture is not of critical importance, careful attention to the intake duct shape is essential, because the pressure gradient is unfavourable here (pressure increasing in the direction of flow), and it is in this area that most of the unnecessary losses are likely to occur. The outlet area needs to be sized to produce the required flow speed through the core, and must take account of the pressure at the outlet location. The ducting everywhere should be designed to avoid strongly adverse pressure gradients and local surface discontinuities.

It is important that the cooling air outlet flow should not interfere detrimentally with the external flow. Exhausting the cooling air from a streamlined aperture on the top of the bonnet (hood) appears to be a good solution, at least for sports racing cars. This means that the intake for interior ventilation cannot be in the conventional location, which is at the base of the front screen, otherwise hot air will be ingested.

Sharp-edged decorative intake grilles and screens should not be used, as they cause a considerable increase in drag and loss of cooling efficiency. The intake should be a streamlined aperture. To prevent ingestion of debris, carefully streamlined slats or guide vanes can be used.

Forward facing ram air intakes may look impressive but a well-designed NACA duct can work nearly as well.

REFERENCES

1. Emmelmann H. J., Aerodynamic development and conflicting goals of subcompacts – outlined on the Opel Corsa, *International Symposium Vehicle Aerodynamics*, Wolfsburg, 1982.

2. Flegl H. and Bez U., Aerodynamics – conflict or compliance in vehicle layout? in *Impact of Aerodynamics on Vehicle Design*, *Proc. of the International Association for Vehicle Design: Technological Advances in Vehicle Design*, SP3, ed. Dorgham M. A., 1983, pp. 9–43.
3. Garrone A. and Masoero M., Car underside, upperbody and engine cooling system interactions, and their contributions to aerodynamic drag, SAE 860212, 1986.
4. Hoerner S. F., *Fluid Dynamic Drag*, Hoerner Fluid Dynamics, PO Box 342, Brick Town N.J. 08723, 1965.
5. Hucho W. H., The aerodynamic design of cars, in Sovran G., Morel T. and Mason W. T. (eds): *Aerodynamic Drag, Mechanisms of Bluff Bodies and Road Vehicles*, Plenum Press, New York, 1978, pp. 313–55.
6. Oler J. W., Roseberry C. M., Jordan D. P. and Williams J. E., paper 910309, in *Vehicle Aerodynamics*, *Recent Progress*, SAE SP-855.
7. Paish M. G. and Stapleford W. R., A study to improve the aerodynamics of vehicle cooling systems, MIRA reports Nos. 1966/15, 1966 and 1968/4, 1967.
8. Renn V. and Gilhaus A., Aerodynamics of vehicle cooling systems, *Proc. 6th Colloquium on Industrial Aerodynamics*, Fachhochschule Aachen, 1985, pp. 303–11.
9. Rivers D. A., Poulter J. E. and Lamont P. J., Aerodynamics of inclined radiators on grand prix cars, *Proc. Vehicle Aerodynamics Conference*, RAeS, Loughborough, July 1994, pp. 13.1–9.
10. Roussillon G., Aerodynamical study of the WM racing car GTP, *Proc. 4th Colloquium on Industrial Aerodynamics*, Fachhochschule Aachen, June 1980, pp. 247–54.
11. Smith C., *Tune to Win*, Aero Publishers, Fallbrook Calif., USA, 1978.
12. Stapleford W. R., Aerodynamic improvements to the body and cooling system of a typical small saloon car, *Proc. 4th Colloquium on Industrial Aerodynamics*, Fachhochschule Aachen, June 1980, pp. 121–32.
13. White R. G. S., An experimental survey of vehicle aerodynamic characteristics, MIRA Reports nos. 1962/1, 1964/4 and 1965/13.
14. Williams J. E., Oler J. W., Hacket J. E. and Hammar L., Water flow simulation of automotive underhood airflow phenomena, paper 910307, in *Vehicle Aerodynamics*, *Recent Progress*, SAE SP-855.

CHAPTER 8

INTERNAL COMFORT – VENTILATION, HEATING, AIR CONDITIONING AND NOISE

With the increasing convergence of popular car designs in terms of performance quality and styling, internal comfort and refinement are now seen as important selling points. Commercial vehicle comfort is also now recognized as being not a mere luxury, but a significant factor in road safety, as it affects the driver's rate of fatigue and ability to concentrate. The effect of internal and external air flows on the comfort level is therefore an important aspect of road vehicle aerodynamic design.

AIR FRESHNESS

Air freshness is principally a matter of the balance between the oxygen and carbon dioxide levels, the latter being produced by the occupants' breathing, but subjective assessment of air freshness is also influenced by the build-up of odours emanating from the occupants, from the trimming materials, and nowadays, less commonly, from engine vapours. Although the carbon dioxide does not pose a direct toxic hazard, if levels become too high, the driver will be unable to concentrate. The stale air therefore has to be replaced continuously. The recommended rate is at least $30\,m^3/h$ ($1060\,ft^3/h$) per person for a fully occupied car, so for a five-seat car, at least $150\,m^3/h$ ($5300\,ft^3/h$) is required. Ventilation rates are often measured in terms of the number of air changes per hour. Thus if the above car had an internal volume of $3\,m^3$ ($106\,ft^3/h$), the number of air changes per hour would be 50. To get some idea what that means in terms of flow rates, $150\,m^3/h$ equates to a jet of air leaving four 6 cm (2.4 in) diameter holes at 4 m/s (13 ft/s). In practice, in order to cope with cooling requirements in extremely hot weather, or heating in near arctic conditions, the required air throughput on a typical car can be more than twice the above figure. Clearly, with such rates of flow, the air vents need to be very carefully positioned if excessive draughts are to be avoided.

Experimental measurement of internal flow velocities is time-consuming and difficult. The internal shape of motor vehicles is complex with many obstructions, making automatic traversing of instruments awkward; the measuring instruments also have to be able to cope with flow angles that vary widely in direction from point to point. The use of computational fluid dynamics (CFD) to model the flow numerically is an attractive alternative to experimentation, particularly as the results do not need to be known with any great accuracy. Ishihara *et al.*[5] describe experimental interior flow measurements using a laser light sheet method, and make comparisons with computed flows.

FRESH AIR INTAKE AND EXHAUST

A popular location for the fresh air intake in domestic cars is just in front of the base of the windscreen, an area where the static pressure is relatively high. The only disadvantage of this location is that if the engine compartment is not well sealed, hot oil smells can be ingested. In general, the intake should be located well above the road level to avoid the ingestion of heavy vapours, and air heated by proximity to the ground.

In the days of coachbuilt cars, there was little need to provide an air outlet, as there were many gaps in the bodywork, but with the high standard of sealing now achieved, it is necessary to include outlet vents. A great deal of design ingenuity is involved in either hiding or making a styling feature of the various inlet and outlet apertures. At one time the outlets were located in areas of low pressure so as to encourage a strong natural pressure-driven flow. The problem with this arrangement, however, is that the air flow rate through the car varies with speed, and the driver has to keep adjusting the aperture of the ventilation openings. In some modern cars, therefore, the outlet is sited so as to produce only a moderate pressure differential between inlet and outlet, and adequate ventilation often requires a small amount of fan assistance.

HEATING

There is no clear agreement as to what represents a comfortable range of air temperatures inside a car. Thermal comfort is to some extent subjective and there are wide differences between what individuals consider to be ideal. In subjective tests, opinions on the optimum temperature vary by some 12° C (22° F) (see Miura[6]). Figure 8.1 gives a rough guide to the range of internal car temperatures regarded as comfortable, and is based on several sources. It will be seen that the range of subjectively comfortable temperatures varies with the external conditions. People acclimatize to seasonal temperature variations, and also alter the thickness of clothing accordingly; thus a slightly higher internal air temperature will be required on a warm summer's day than in spring. In very cold exterior conditions, the internal air temperature must be raised in order to offset the effect of rapid heat loss due to radiation through the windows. The DIN standard (DIN 1946) recommends 22° C when the external ambient temperature is in the range −18 to +20° C.

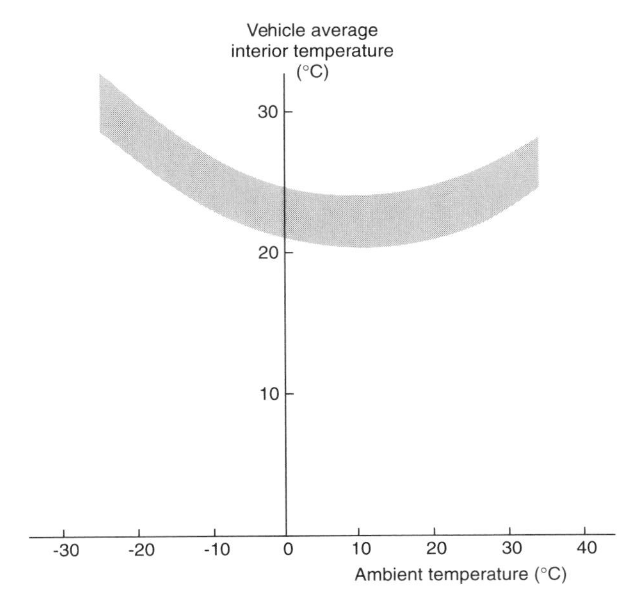

Fig. 8.1 Range of comfortable internal temperatures as a function of the external ambient temperature.
(*Based on data from a number of sources*)

HEATER SYSTEM DESIGN REQUIREMENTS

Internal heating is provided by drawing in cold external air, and passing it through a heat exchanger which normally resembles a small engine cooling radiator. The heater has to be able to heat the air to a comfortable temperature at the flow rate required for ventilation, when the external temperature is at the lowest value likely to be encountered. In cold northern climates this can necessitate a large heat exchanger unit.

Humans feel comfortable when the air temperature around their feet is roughly 7° C (13° F) higher than that around their head, except in very hot conditions. The heater system therefore has to be able to supply hotter air to the feet than to the upper body area. The temperature stratification (variation of temperature with height) is provided by passing a proportion of the fresh air through the heater, and mixing this with unheated air in different proportions for the face/body-level vents and footwell vents. In older cars, the face-level vents only supplied cold air, but in more modern systems, a variable proportion of hot air can be mixed in. In addition to warming the occupants, the heater has to be able to supply hot air to defrost and demist the windscreen, and this entails adding another set of ducting and associated controls.

In most cars, the air flow rate is adjusted by changing the fan speed. Distribution of heated air between the demist and footwell vents is controlled by a lever or knob.

Upper body vent apertures and the direction of the outlet air streams are normally operated by individual controls at each vent outlet.

On older systems, the air temperature was altered by varying the flow rate of water through the heat exchanger. An alternative more precise and rapidly responding temperature control is provided by having two streams of air. One stream is passed through the heater, while the other bypasses it. The hot and cold air streams are then mixed in the required proportions. One slight disadvantage of this method is that unless the sealing and separation of the hot and cold streams is very good, the air may still be slightly heated even when set to pass only unheated air in the 'cold' position. In very hot weather, even a 2° C rise in temperature can be the last straw in terms of discomfort.

DRAUGHT COOLING

At high outside temperatures, or with a high input of radiant heat through the glass, some cooling effect can be provided by creating an air stream around the occupants. The air flow removes heat by conduction, and in very hot conditions helps to evaporate sweat. The latent heat of evaporation increases the heat transfer rate from the body, and produces a cooling effect even when the ambient air temperature is higher than the body temperature. Excessively high air velocities, however, lead to discomfort, particularly in exposed areas such as the face and neck, where a comfortable maximum is around 3 m/s. Higher velocities are acceptable at chest level. The ventilation system should therefore be designed to avoid draughts: local areas of high velocity. Most vehicles have chest-level outlet vents which can be articulated so as to direct the air stream in the desired direction. When the external air is both hot and humid, the cooling effectiveness of an air stream is reduced, as the moist air inhibits sweat evaporation. Under these conditions some artificial cooling or air-conditioning becomes necessary.

HUMIDITY

Humidity relates to the amount of water vapour held in the air. The amount that the air can retain increases with the air temperature. Relative humidity is the amount of water vapour present relative to the maximum amount that can be held at that temperature. At 100 per cent humidity, the air is said to be saturated. The level of humidity required for human comfort is in the range 30–70 per cent.

If the humidity level falls below about 30 per cent, some discomfort may be felt, particularly in terms of dryness of the throat. Very low humidity levels can occur in the winter when cold external air is heated. The same is true in airliners, where the external air temperature is around −55° C during cruise. In buildings low humidity is sometimes corrected by adding water vapour, but in cars this is considered impractical and unnecessary. Low levels of humidity seem to be tolerable even for the very long journey times involved in intercontinental air travel, as long as sufficient liquid refreshment is available, and low humidity is therefore not regarded as a major problem in car comfort. By contrast, if the humidity rises above 70 per cent, the feeling becomes oppressive, as before a thunderstorm, even if the temperature is low; asthma and hay-fever sufferers are affected. In hot weather high humidity is very unpleasant, and an air-conditioning system provides the solution.

AIR-CONDITIONING

Air-conditioning units involve the use of a refrigerating heat exchanger which cools the incoming air. These units are now standard on vehicles in many parts of the world including much of North America, but are less common in Europe. Apart from producing a comfortable temperature, cooling the air provides the ability to reduce the humidity. If very humid warm air is cooled sufficiently, some of the water will condense out. To reduce the humidity of the air, therefore, it is cooled below the finally required temperature, typically to −10° C. The water vapour that condenses out is drained off, and the air is then warmed to the required temperature, either by passing it over another heat exchanger, or by mixing it with warmer air. As the temperature rises, the relative humidity falls.

Apart from the cost and complexity of air-conditioning systems, a major disadvantage is that around 4 kW (5.4 bhp) of engine power is absorbed in driving the pump of the refrigeration system. This represents a significant proportion of the average engine output power, particularly on small cars, and results in poor fuel economy and reduced performance when the system is running.

From the foregoing, it can be seen that in current vehicles, there is a complex arrangement of internal flow systems and ducting. Poorly designed systems will not only affect the interior comfort and noise levels, but can also influence the overall efficiency of the vehicle by making excessive demands for power.

AERODYNAMIC NOISE

Another important aerodynamic influence on comfort and refinement is the noise generated by both the internal and external air flows. Two aspects of noise are of concern to the vehicle designer: interior noise, which affects the driver and passengers, and drive-by noise, which is the noise heard by a bystander.

Acoustic noise is the effect of rapid pressure changes transmitted as waves travelling at the speed of sound. The intensity of the sound that is heard is dependent on the mean square (the variance) of the pressure fluctuations. In practice, it is usually the square root of the mean square (the rms) of the pressure fluctuations that is measured, and this is known as the sound pressure.

The human ear has a non-linear response to sound intensity, and a doubling of the intensity is not perceived as being twice as loud. For this reason a logarithmic measurement scale called the decibel (dB) is normally used, and the acoustic noise level is usually expressed in terms of the sound pressure level (SPL), which is defined by

$$\text{SPL (in dB)} = 10\log\left(\frac{\text{sound intensity}}{\text{reference sound intensity}}\right)$$

which is identical to

$$20\log\left(\frac{\text{sound pressure}}{\text{reference sound pressure}}\right)$$

On the decibel scale, a doubling of the sound intensity will produce an increase of 3 dB since log 2 is approximately equal to 0.3. An increase from 74 dB to 77 dB therefore represents a doubling of the intensity.

SOUND SPECTRA

Noise, as opposed to pure musical notes, consists of sound spread over a range of frequencies. Apart from measuring the overall sound intensity, it is important to know how that intensity varies with the frequency, and this is provided by a noise spectrum. In a conventional noise spectrum, the frequency range is split into octaves, and each octave is split again into three parts. The sound pressure level in decibels is measured over each $\frac{1}{3}$ of an octave frequency range, and plotted against the frequency in the centre of that range (see Figs. 8.2 and 8.4).

The human ear only detects sound in the range of approximately 20 Hz to 20 000 Hz, so it is normal to use a noise level scale which is weighted towards sound in the most readily audible frequencies; that is, sound measurements in the well-audible frequency range are multiplied by a factor greater than 1, whereas at the ends of the range, the values are reduced. This weighted scale is denoted by dB(A).

EXTERNAL DRIVE-BY NOISE

In many countries, there is legislation aimed at reducing vehicle drive-by noise, and standard tests are prescribed. According to Hardy,[3] the impending new European standard of 74 dB(A) for drive-by noise tests may be difficult to meet for some vehicles, as the tyre noise alone is nearly at this level; modern wide tyres tend to be noisier than the older narrow forms.

Most types of aerodynamic noise rise as the sixth power of the speed (see Haruna *et al.*[4]), and since the critical drive-by tests involve relatively low speeds in low gears, aerodynamic noise is not yet a major factor in terms of meeting legal requirements; however, as the requirements are made increasingly stringent, and as engine silencing improves, it may become significant. At present, aerodynamic noise is seen mainly as a problem of the internal environment of the vehicle.

INTERNAL NOISE PROBLEMS

Internal noise derives partly from internal sources such as the ventilation system, and partly from the effects of the external air flow. The noise levels due to external sources are generally the more important, particularly at high speed.

There are two approaches to the problem of reducing externally generated noise: removing the sources, and isolating the interior of the vehicle from the exterior sound source. Direct transmission through body panels may be reduced by lining the panels with vibration-damping materials, but the major transmission path is usually through leaks, so careful attention to sealing is very important. The slightest opening of an aperture such as a window can cause a large increase in noise transmission to the interior. Gaps around doors etc. are a source of sound generation as well as forming noise transmission paths.

AERODYNAMIC NOISE MECHANISMS

Most aerodynamic noise is generated by pressure fluctuations associated with turbulence and vortical structures. It is therefore fortunate, and not merely coincidental, that design features aimed at reducing aerodynamic drag also reduce the sources of external noise. The low-drag shapes now used, together with a high standard of sealing, mean that the internal noise of domestic cars has dropped significantly from the levels that were common up to the early 1980s. According to George[2] internal noise levels in the range 70–80 dB(A) are tiring on long journeys. In the 1970s, Buchheim *et al.*[1] looked at 15 different vehicles, and found typical internal noise levels in the range 62–78 dB(A) at 113 km/h (70 mph), rising to 72–87 dB(A) at 180 km/h (110 mph). Piatek[7] quotes internal noise figures for a small car driving at 150 km/h as:

- 82.5 dB(A) from the engine
- 78 dB(A) from the tyres
- 78.5 dB(A) due to external air flow

which equates to a total of 85 dB(A).

BOUNDARY LAYER NOISE

The noise generated by a turbulent boundary layer is somewhat random in nature, and is spread over a wide frequency range, as may be seen in Fig 8.2. In cars it is difficult to distinguish this source from all the other noise inputs, but in a glider it can be heard clearly as a rushing sound with no distinct tone. In a car, noise due to the boundary layer turbulence is not usually the major source, and much of it is at high frequencies which are relatively easy to isolate by the application of absorbent materials to the inside of body panels, and by good sealing.

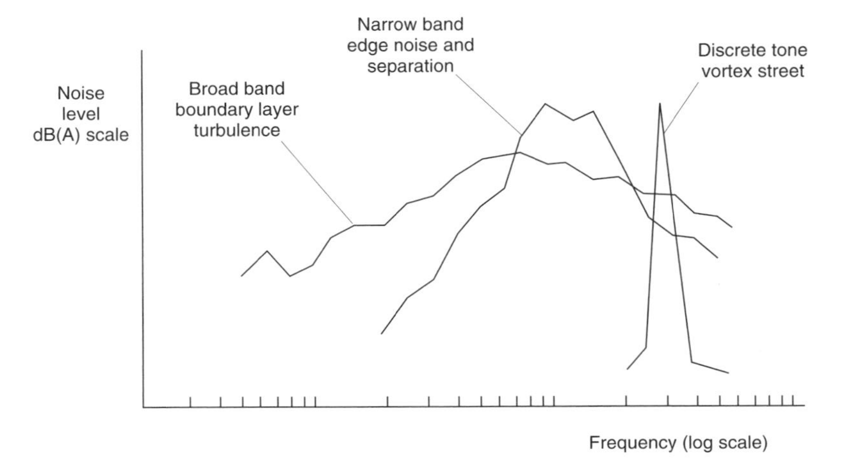

Fig. 8.2 Aerodynamic noise spectra associated with different flow mechanisms.

EDGE NOISE

Flow separating from sharp corners and the edges of styling details usually produces noise that is distributed over a fairly narrow frequency band, as indicated in Fig. 8.2. As illustrated in Fig. 8.3, the separated shear layer emanating from an edge contains a high level of vorticity, and tends to roll up into large eddies. The large eddies break off at semi-regular intervals, and in addition, small sub-eddies are formed. The process is not normally quite regular enough to produce a single tone, although under resonant conditions, such as being adjacent to a resonant cavity, this can happen; indeed it is the mechanism of sound production in a flute.

Despite the fact that the noise has no single tone, the ear can detect whether the band of frequencies is steady, or moving up the scale. The frequency of the

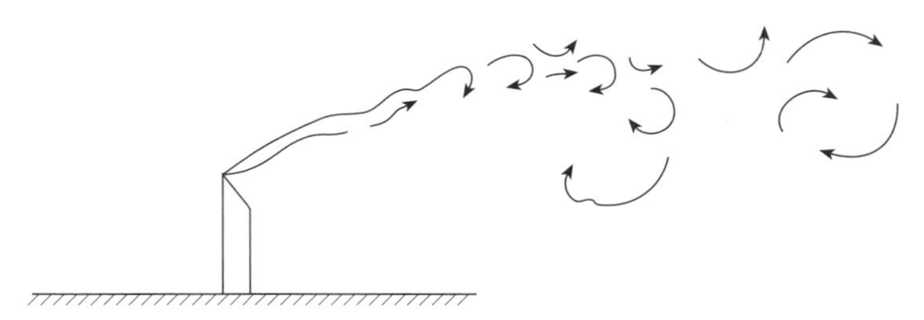

Fig. 8.3 Flow separation from a sharp edge rolls up into large vortices, which also break into smaller ones.

centre of the band increases with air speed, and this gives the occupants an audible indication of the driving speed. Edge separation noise can be reduced by making the surface of the vehicle as smooth and continuous as possible. Gaps between panels should be small or flush sealed, and there should be no protrusions. Sharp corners should also be avoided. Fortunately these are also the requirements for low drag, so a good low-drag design should also be relatively quiet.

One of the largest sources of wind noise is the strong vortex that usually forms at the A-posts. This source of noise has been studied extensively (Haruna *et al.*[4] and George[2]). The vortex forms over the front glass sidelight, which is a reasonably good sound transmitter, and the noise is contained in a relatively small frequency band. Making the A-posts well-rounded with a continuous contour can help reduce the vortex strength, but a very large radius is incompatible with visibility requirements.

Two items that often generate noise unnecessarily are wing mirrors and wheel trims. It is perfectly possible to produce stylish designs for these elements without compromising the noise and drag requirements.

Figure 8.4 shows measured spectra for different types of local flow obtained in wind-tunnel tests by Stapleford and Carr.[8] It will be seen that attached boundary layer flow only becomes significant at the higher frequencies, whereas separated flow is important at low frequencies. By far the largest contribution is that due to vortex flow.

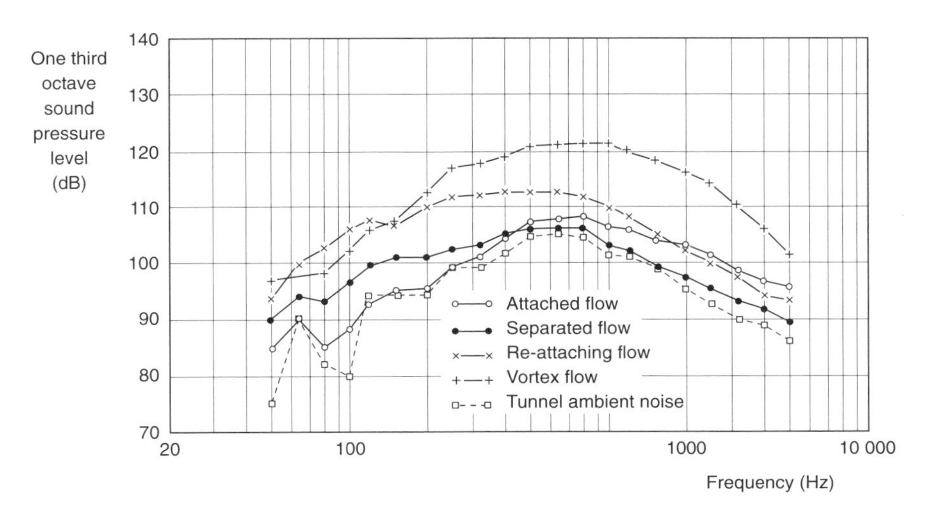

Fig. 8.4 Spectra for various types of local flow around a car, measured in a wind-tunnel. The reference pressure for the dB scale was 2×10^{-5} N/m^2.
(*After Stapleford and Carr,*[8] *courtesy of MIRA*)

DISCRETE TONES DUE TO VORTEX SHEDDING

The most annoying type of noise is that of discrete tones. Such tones are generated by the production of a Karman vortex street behind long bluff shapes, a good example being the elements of a roof rack. As described in Chapter 1, bluff objects shed vortices alternately from the two sides at a frequency f that is related to the air speed U and the depth d by the expression

$$S = fd/U$$

where S is called the Strouhal number and has a value of 0.146 for a long flat sheet and 0.2 for a long circular rod.

The value of the Strouhal number changes with aspect ratio, and stubby objects with low aspect ratio produce more disordered vortex patterns and consequently less of a discrete tone. Roof racks which are composed of long thin elements with high aspect ratio can produce a very loud and unpleasant monotone. Using a Strouhal number of 0.2 for a circular rod, it can be readily calculated that at 113 km/h (70 mph, 32 m/s) the vortex shedding frequency from 1 cm diameter circular rod elements of a roof rack would be 640 Hz; not far from the nicely audible 440 Hz musical 'A' used as a tuning note for orchestras. The use of a streamlined rooftop container instead of a bare rack alleviates the problem.

Fans are another source of discrete tones, the worst offender in this respect being the radiator cooling fan. The fan blades are effectively rotating wings, and therefore shed trailing vortices. These are wound into a helix downstream of the fan, and as the helix cuts through obstacles, the resulting pressure disturbances occur at discrete frequencies. In addition, there is a large amount of semi-random noise due to turbulence. To inhibit the generation of discrete tones, fans are usually made with uneven spacing between the blades, and often with an uneven number of blades.

Fans were a major contributor to the engine-related noise in older vehicles, where they were driven directly by a belt drive from the engine, and thus the fan noise increased with engine speed. As described earlier, the fan is now nearly always driven electrically, and only switches on when the coolant temperature exceeds a certain limit. The introduction of electrically driven fans accounts for much of the reduction of vehicle internal noise that has occurred since the 1970s.

LOW-FREQUENCY SUB-AUDIO NOISE

Aerodynamic effects can generate pressure fluctuations that are too low in frequency to be audible, but can be felt, and can be very unpleasant. One major source is the sun-roof aperture that has regained popularity on closed cars. The aperture acts rather like the mouthpiece of a very large but inaudible flute, with the car interior acting as a resonator. Similar effects can occur when windows are opened. In the case of the sun-roof, the solution is to raise a small deflector at the

Fig. 8.5 The automatically deployed deflector at the front of the sun-roof aperture on a Rover 827.

front of the aperture when the sun-roof is opened. This forces a strong flow separation which inhibits intermittent reattachment on the rear of the aperture. Figure 8.5 shows the deflector on a Rover 800 series saloon; the deflector rises automatically as the roof slides backwards.

DESIGN IMPLICATIONS

Internal flows for ventilation and the control of occupant comfort are now a major aspect of vehicle design, both for domestic cars and large commercial vehicles, and the complexity of the systems required means that they cannot be added as an afterthought. The siting and styling of inlet and outlet apertures is a significant consideration.

Noise is another important aspect of vehicle aerodynamics, but since most aerodynamic noise is generated by turbulent or vortical structures, the requirements of low drag and streamlining are happily consistent with noise minimization.

REFERENCES

1. Buchheim R. W., Dobrzynski H., Mankau H. and Schwabe D., Vehicle interior noise related to external aerodynamics, *International Journal of Vehicle Design*, Special publication SP3, 1968, pp. 197–209.
2. George A. R., Automobile aerodynamic noise, SAE paper No. 900515, 1990.
3. Hardy K., Tyre noise – the villain, in *Noise and the Automobile*, selected papers from Autotech 93, Birmingham, 16–19 November 1993, MEP, London, pp. 117–21.
4. Haruna S., Kamimoto I. and Okamoto S., Estimation method for automobile noise, SAE paper No. 920205, in SAE SP 908, 1992.

5. Ishihara Y., Shibata M., Hoshino H. and Hara J., Analysis of a full-scale passenger-compartment model using a laser-light-sheet method, SAE paper No. 920206, in SAE SP 908, 1992.
6. Miura T., Studies on the optimum temperature. On some factors affecting the shift of optimum temperature, *Journal of Science and Labour*, Vol. 44, 1968, pp. 431–53.
7. Piatek R., Operation safety and comfort: in Aerodynamics of Road Vehicles, ed. Hucho W., Butterworth, London, 1986.
8. Stapleford W. R. and Carr G. W., Aerodynamic noise in road vehicles, part 1:- the relationship between aerodynamic noise and the nature of airflow, MIRA rep. No. 1971/2, 1970; Stapleford W. R., Aerodynamic noise in road vehicles, part 2: A study of the sources and significance of aerodynamic noise in saloon cars, MIRA rep. No. 1972/6.

CHAPTER 9

OPEN CABRIOLET VEHICLES

With the introduction of chassisless construction in the early post-war period, the drophead or convertible car fell out of favour, at least in the eyes of most manufacturers. Without a chassis it is difficult to retain adequate stiffness in the structure when the top of the shell is removed. Reinforcement has to be added, which tends to make the cabriolet version heavier than the saloon, the reverse of what is required for the sporting image. Soft-top cars are also noisier and something of a liability in terms of security, and this restricts the market appeal; nevertheless, manufacturers have recently recognized that there is a sizeable niche market for convertible versions of their closed cars, and even for the traditional open sports car. There are now cabriolet versions of several popular models. At one time owners of such vehicles did not expect much in terms of creature comfort, and indeed being exposed to the elements was perceived as part of the appeal of the open car. Attitudes change, however, and a good internal environment is now seen as being essential to the commercial success of convertible vehicles. It has to be admitted that old vehicles such as the author's Brough Superior drophead (Fig. 9.1)

Fig. 9.1 Brough Superior d.h.c. of 1936
The internal wind environment in classic open touring cars usually required some protection for the occupants, particularly in the rear seats.

Table 9.1 Comparison of C_D values between closed and open-top versions of two vehicles. Note that the Mercedes wind-tunnel model was not an exact replica, and the figures are not the same as for the real vehicle.

Vehicle style	C_D values	
	Rover 200 series (full-size car)	Semi-scale model based approximately on the Mercedes E220
Closed saloon vehicle	0.34–0.36 [depending on trim]	0.36
Cabriolet, soft-top raised (fully developed vehicle)	0.37 (2.8%)	
Open cabriolet baseline side screens down	—	0.51 (41.7%)
Open cabriolet baseline side screens up	—	0.48 (33.3%)
Open cabriolet after optimization, side screens up	0.39 (11.4%)	0.41 (13.9%)

may look inviting on a hot day, but life is quite hard for rear-seat passengers at open-road cruising speed. Even the front-seat occupants need a certain amount of protection from strong head-level draughts and buffet.

Another factor that has to be addressed nowadays is that open variants invariably have a higher drag coefficient than the closed model. Table 9.1 shows a comparison between the drag coefficient values for two sets of vehicles, the Rover 200 series, and a wind-tunnel model based approximately on the shape of a Mercedes E220. From the data for the model, it can be seen that simply removing the hard-top can produce a large drag penalty, however, with the corrective measures described later, and some development work, it is possible to reduce the penalty to around 12 per cent or less. The data for the Rover show that a high drag penalty is not inevitable.

SOFT- AND HARD-TOPS

The effect of lowering the soft-top on a convertible vehicle is not quite the same as converting from a metal hard-top to an open configuration, as may be seen from the Rover data. The raised soft-top does not in general have such smooth contours as the hard-top, and has a higher drag coefficient.

In a survey of five vehicles of the 1980s, the author[1] found increases in drag coefficient due to lowering the soft-top to be in the range 18–26 per cent, with a

Table 9.2 Comparative C_D values for various vehicles with the soft-top in raised and lowered positions.

Car	C_D values		
	Top raised	Top down	Percentage increase
BMW Z1	0.36	0.43	19
BMW 325i	0.38	0.48	26
Corvette	0.38	0.45	18
Mercedes SL	0.34	0.41	20
Porsche 911	0.39	0.49	25
Rover 216	0.37	0.39	5.4
Beetle	0.5	0.68	36

Table 9.3 Increase in drag coefficient due to lowering the soft-top on a cabriolet vehicle.

Vehicle style	Increase in C_D	
	Side screens up	Side screens down
4 seater	0.025–0.07	0.05–0.1
2 seater	0.025–0.085	0.025–0.097

(Based on data in Cogotti[4])

mean of 21.6 per cent. These correspond roughly to increases in C_D in the range 0.063 to 0.091. Going back somewhat further into history, opening the old VW Beetle cabriolet produced a drag increase of 36 per cent. By contrast, the more recent Rover 216 shows only a 5.4 per cent increase ($\Delta C_D = 0.019$). These data are shown in Table 9.2. Cogotti[4] gives increases in C_D in the ranges shown in Table 9.3.

Another factor that needs to be taken into account is that soft-tops tend to balloon at high speed. Considerable distortion in shape can occur if the material is too flexible, or inadequately attached to the supporting members.

THE SOURCES OF THE DRAG INCREASES IN OPEN CARS – THE FLOW STRUCTURE

The fact that there is such a large degree of variability in drag coefficient between vehicles which may be quite similar in appearance shows that the drag penalty is sensitive to small design differences. It therefore follows that it is possible to minimize the drag rise by understanding the drag-producing mechanisms, and making appropriate design changes.

The flow structure in open cars is dominated by a large separation bubble that forms in the driver/passenger area. As illustrated in Fig. 9.2, this separation bubble is bounded by a separated shear layer originating from the front screen. Experimental investigations by the author (Barnard[2]; Barnard and Monaghan[3]) show that the bubble tends to be unstable, as may be seen in the flow visualization photographs of Fig. 9.3. In the first picture, the shear layer can be seen to be partially reattaching on the boot (trunk) lid. The bubble appears to grow slightly, until it bursts or vents, and then reattaches again to repeat the cycle. There must be a great deal of irrecoverable turbulent energy associated with this large unstable structure, and it therefore represents a major source of drag. Cogotti[4] shows detailed wake studies comparing the closed and open-top cases. The open vehicle shows a much larger wake.

Other features of the flow are illustrated in Fig. 9.2. The low-pressure area under the bubble draws in air from the sides, and strong vortices form at the A-posts. The vortices and turbulent mixing add further increments to the drag. With the side screens raised, the inflows from the side are inhibited, thereby reducing the turbulent interactions; also, the A-post vortex can be weakened by flow attachment on the screens. The main separation bubble becomes more two-dimensional and ordered, and consequently, as may be seen in Table 9.3, the drag is reduced.

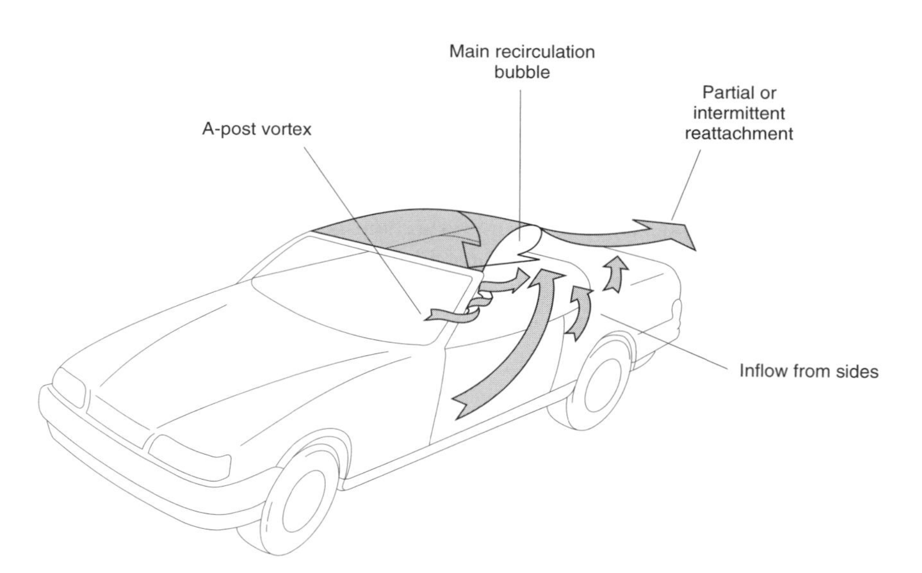

Fig. 9.2 Features of the flow in an open cabriolet
The flow structure is dominated by a large separation bubble which normally reattaches partially on the boot (trunk) lid.

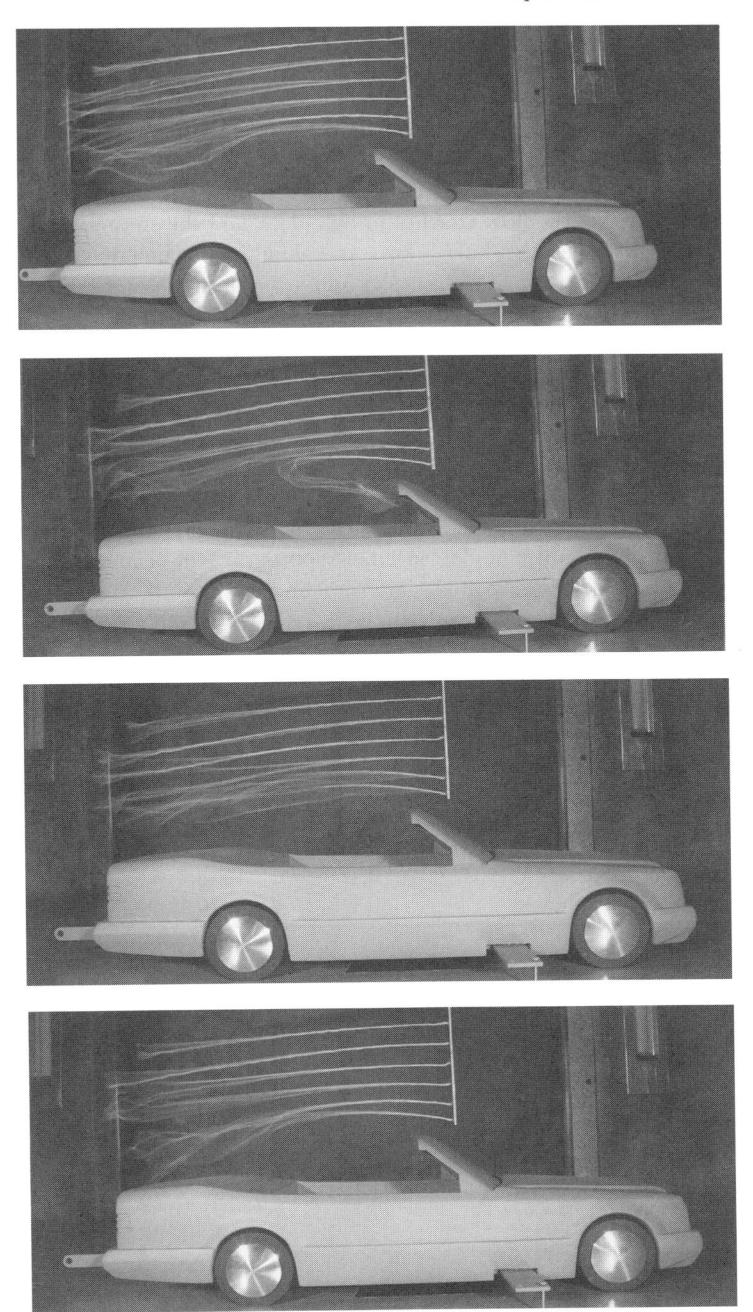

Fig. 9.3 Flow visualization of the unsteady separation bubble in an open cabriolet
Reattachment at the rear is partial and intermittent. Note the forward flow evident in one of these pictures. The forward flow comes surprisingly close to the bounding separated shear layer.

DESIGN FEATURES FOR REDUCING THE DRAG

In order to reduce drag it is necessary to stabilize and reduce the size of the separation bubble, and to encourage reattachment. One of the most effective methods of stabilization is to introduce a deflector or guide plate to the front-screen header-rail, as illustrated in Fig. 9.4. Experimental studies (Barnard[1,2]; Barnard and Monaghan[3]) show that the flow attaches to this plate, and the shear layer is pulled down, reducing the separation bubble size, and aiding reattachment at the rear. The deflector plate can conveniently double as a sun visor, an important requirement in an open car, where the driver's eyes are often exposed directly to the sun.

A roll-over bar can have a similar influence, but is less effective. Where such a roll-over bar is incorporated, its positioning and cross-sectional profile should be optimized for aerodynamic effectiveness using wind-tunnel testing. Table 9.4 shows the influence of introducing a header-rail deflector and a roll-over bar on the semi-scale Mercedes model.

Increasing the screen rake helps to control and tighten the separation bubble, and although it may be impractical in manufacturing terms to alter the screen rake for a cabriolet variant of a popular model, an 'open' sports car will benefit from a high screen rake angle. Extreme rake angles are sometimes employed, as may be seen in Fig. 9.5, although this can cause problems with reflections, and solar heating.

Table 9.4 The effect of a header-rail and a shaped roll-over bar on the drag coefficient of the semi-scale Mercedes model.

Vehicle style	C_D	Percentage reduction in drag
Baseline cabriolet – side windows up	0.46	—
With shaped roll-over bar	0.43	6.5
With header-rail deflector	0.42	8.7
With roll-over bar and deflector	0.412	10.4

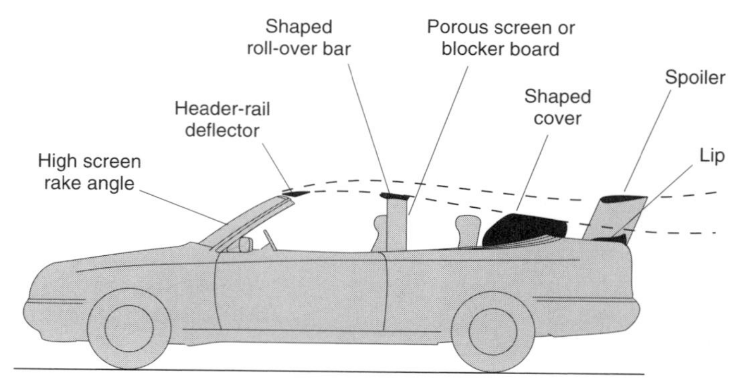

Fig. 9.4 Some means of controlling and stabilizing the separation bubble. The most effective device is the header-rail deflector plate.

Fig. 9.5 High screen rake angle on the Lotus helps to tighten and stabilize the separation bubble.

To promote reattachment on to the rear bodywork, it should be raised as much as possible consistent with styling requirements and rearward visibility. If the folded top is wrapped in a soft cover, suitable shaping and siting of the cover can help the reattachment process.

THE INTERNAL ENVIRONMENT IN OPEN CARS

At one time the very idea of fitting a foot-well heater to an open sports car would have been considered laughable, but in an attempt to appeal to a larger market than the young and macho, manufacturers now emphasize interior comfort and absence of draughts in their advertising promotions.

The major source of discomfort in the open car is the separation bubble, and the associated inflows. The discomfort is associated with

- draughts
- excessive cooling
- buffet

Buffet can be described as a combination of low-frequency pressure fluctuations, and rapid changes and reversals of local internal flow speed and direction. Even a modest degree of buffet can be tiring and unpleasant after a few minutes. Design features, such as a header-rail, that help to stabilize the vortex will reduce the buffet effect.

Rather surprisingly, as indicated in Fig. 9.2, the flow inside the passenger area is predominantly reversed, that is, the flow direction is from the rear towards the front. If the driver's hat falls off, it is more likely to blow forward and end up in the foot-well than to blow backwards. The forward flow can produce uncomfortable draughts, particularly around the head and neck. Various techniques can be employed to mitigate this effect, the simplest being to use the seat as a wind-shield. The seats need to be wide enough to cover the shoulders of the occupants, and a solid integral head-rest/restraint can be be used to keep draughts away from the

Fig. 9.6 The Aston Martin Virage Volante
Tests on the prototype vehicle showed a good internal environment. Features that contribute to this are the generous seats, wide doors, well-raked screen, and the soft-top cover which provides some shielding, and a convenient line of reattachment. The hood cover seen in the photograph was in fact a temporary mock-up. The final line was lower.
(*Photograph courtesy of Aston Martin Lagonda Ltd*)

neck. Up-market vehicles such as the Aston Martin Virage Volante (Fig. 9.6) have deeply upholstered seats that shield the occupants effectively even at 144 km/h (90 mph) (Barnard and Monaghan[3]).

Draughts from the side can be largely prevented by raising the side screens. With these screens lowered, it is an advantage to have doors that are wide and high in relation to the seating. In this respect, relatively large vehicles such as the Aston Martin have a definite advantage compared with smaller sports cars.

DRAUGHT SCREENS

A feature that is found on one or two open cabriolets is the draught screen, which goes under a variety of proprietary names. This consists of a board or screen placed behind the front seats of some two-seat or close-coupled 2 + 2 configured vehicles. The purpose is to prevent the draughts associated with the forward flow of air. A shoulder-height solid board fitted experimentally to the prototype Aston Martin Virage Volante (Barnard and Monaghan[3]) produced a 91.5 per cent reduction in the mean draught velocity in the near vicinity of the front occupants, but the board was not offered on production vehicles, as the wide well-contoured seats of the production vehicle proved to be more than adequate for preventing draughts. One possible method of providing further protection would be to draw up a hinged lid from the central console to provide infilling between the front seats.

Even greater draught protection is provided by a full head-height transparent screen as seen fitted to the Porsche cabriolet in Fig. 9.7. The screen is made from a porous flexible material with a high level of transparency; at a distance, it resembles a slightly tinted glass screen. The screen can be lowered when not

Fig. 9.7 This Porsche features a retractable porous screen that provides the front seat occupants with good protection from draughts. The mechanism also includes a panel which rises to seal the gap between the screen and the rear of the cockpit.
(*Photograph courtesy of Porsche AG*)

required. When raised, it incorporates a panel which effectively seals the gap between the screen and the rear bodywork. This ingenious arrangement is highly effective in preventing draughts. There are obvious problems in adapting the idea to a full four-seat configuration vehicle, although parents of small children might welcome the idea of being able to partially isolate the front and rear occupants. With some ingenuity, the screen could be coupled with a roll-over bar.

As mentioned earlier, when the soft-top is folded down, its cover can be used to form a point of flow reattachment. By means of wind-tunnel tests (Barnard and Monaghan[3]), it is possible to optimize the profile of the cover to aid reattachment, and thereby improve both the internal wind environment and the drag coefficient; both height and positioning are important.

NOISE AND OTHER ASPECTS

Opening a vehicle means that the occupants are exposed to the full external noise environment, but this is to some extent compensated for by the removal of resonances. Externally transmitted engine noise is nowadays quite low, and with good low-drag design, the aerodynamic noise is not excessive at normal cruising speeds. The main problems are tyre noise, the effects of the A-post vortices, and noise associated with the separated shear layer.

The instability of the shear layer is at a low sub-audio frequency that is felt rather than heard, and is part of the buffet effect. A modern low-drag cabriolet optimized using the methods described here and conforming to accepted standards of engine silencing should not be excessively noisy.

With the soft-top raised, the transmission of noise is still much greater than for a solid metal top. The flexibility of the material allows low-frequency noise to be transmitted with little loss, but high frequencies are reasonably well attenuated in high-quality lined and padded soft-tops. Unfortunately, such tops occupy a large amount of space when folded. Careful attention to sealing round the side screens is important, a factor that was almost totally neglected on older designs. Cogotti[4] shows the results of noise measurements in cabriolet vehicles.

One final aspect of open car driving worthy of mention is that despite the high level of ventilation, noxious exhaust gases can be drawn back into the interior by the low pressure therein. The location of the exhaust outlet is therefore important.

THE RELATIONSHIP BETWEEN DRAG AND COMFORT

Since the unstable separation bubble is the source of both the high drag and the internal environment problems in open cars, it follows that measures taken to improve the internal environment will usually result in a reduction in drag: an assertion that is borne out by the results of the experiments described in Barnard[2] and Barnard and Monaghan.[3] Table 9.5 shows the influence on drag of adding various combinations of the features outlined above. The drag reduction due to the

Table 9.5 The effect on the drag coefficient of various design features aimed at stabilizing the separation bubble and improving the internal environment.

Vehicle style	C_D
Hard-top version	0.36
Side windows down	
basic open cabriolet	0.51
with shaped roll-over bar and draught screen	0.48
with header-rail deflector plate	0.46
with deflector plate and roll-over bar and draught screen	0.45
Side windows up	
basic open cabriolet	0.46
with shaped roll-over bar	0.43
with header-rail deflector plate	0.42
with roll-over bar and draught screen	0.415
with roll-over bar and deflector plate	0.412
with roll-over bar, deflector plate and draught screen	0.41

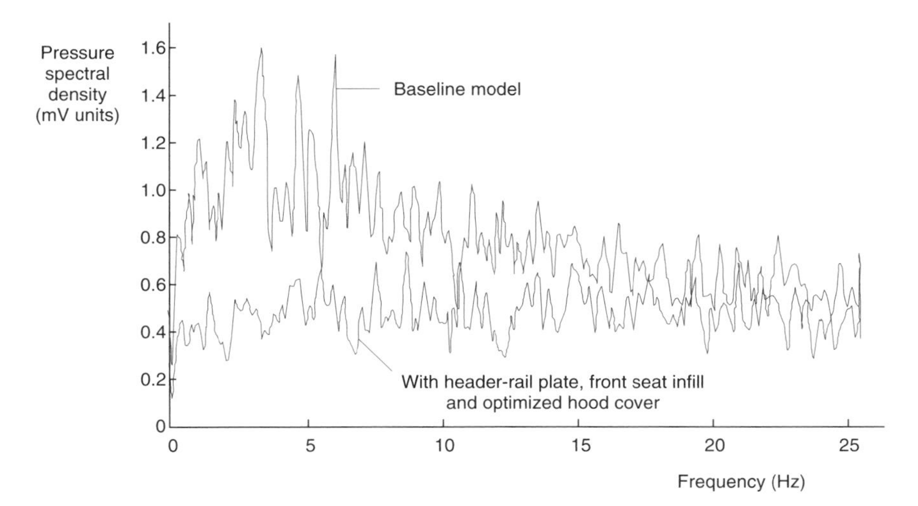

Fig. 9.8 The effect on the spectrum of internal velocity fluctuation of the addition of various features aimed at reducing draughts and controlling the separation bubble. (*Based on data from model tests on the Virage Volante*)

presence of the draught screen is an unexpected bonus. Figure 9.8 shows how a combination of similar features affected the velocity fluctuations measured inside a model of the Virage Volante (see Barnard and Monaghan[3]).

When optimizing an open vehicle, it is important to conduct tests in the presence of cross-winds. During the tests described in Barnard and Monaghan,[3] it was found that strong cross-wind significantly degraded the interior environment. The drag coefficient may also be adversely affected.

DESIGN FEATURES FOR OPEN CABRIOLET VEHICLES

The main requirements for low drag are similar to those for reduced internal draughtiness and buffet, and involve finding means of stabilizing and reducing the extent of the large separation bubble that forms in the passenger compartment. The design features that encourage this are high screen rake, and a raised rear deck.

Additional features that can be incorporated are:

- a deflector plate attached to the front-screen header-rail
- an aerodynamically tailored roll-over bar
- a suitably profiled soft-top cover

All of these features need to be optimized using wind-tunnel tests; it is virtually impossible to produce the required geometry by purely intuitive design.

Draught protection for the occupants can be provided by:

- wide contoured seats with integral head restraints
- an infill board placed behind the front seats to prevent air flow below shoulder height
- a full head-height porous transparent screen which can be erected behind the front seat occupants

When all or most of the above features are incorporated, it is possible to produce a low-drag cabriolet with a surprisingly comfortable internal environment.

REFERENCES

1. Barnard R. H., Aerodynamic problems and solutions in open cabriolet vehicles, *Proc. RAeS Conference: Vehicle Aerodynamics*, Loughborough, July 1994, pp. 2.0–2.7.
2. Barnard R. H., Improving the wind environment in open cars, paper C487/021/94, *Proc. I. Mech. E. Conf. on Vehicle NVH and Refinement*, Birmingham, May 1994.
3. Barnard R. H. and Monaghan P., Assessment and reduction of wind buffeting in an open cabriolet, *International Journal of Vehicle Design*, Vol. 13, nos. 5/6, 1992, pp. 486–93.
4. Cogotti A., Experimental techniques for the aerodynamic development of convertible cars, paper No. 902347, in *Vehicle Aerodynamics: Wake Flows Computational Fluid Dynamics, and Aerodynamic Testing*, SAE SP-908, 1992, pp. 183–201.

CHAPTER 10

VEHICLES IN CROSS-WINDS

WIND HAZARDS FOR VEHICLES

On 25 January 1990 an unexpectedly severe storm hit England. An analysis by Baker and Reynolds[3] shows that during this period there were at least 390 wind-related road accidents involving injury. Of these 47 per cent involved overturning, 19 per cent were due to course deviation, and the rest were mainly attributable to hitting fallen trees and debris. 66 per cent of the accidents involved trucks, buses and light vans, and it is safe to assume that most if not all of the overturning incidents were associated with these vehicles.

In terms of overall accident statistics, the effects of severe storms may not be very important, but Kobayashi and Kitoh[16] report that about 1.2 per cent of all accidents on the Tohoku Expressway in Japan have been found to be due to cross-winds. Although small in percentage terms, this still represents a large number of accidents annually, particularly if the data can be extrapolated to cover the whole country.

Not many wind-induced accidents involve overturning; domestic cars are unlikely to blow over, as we will see later, and accidents are mostly associated with excessive course deviations resulting in impacts with other vehicles or roadside objects. Overturning can occur indirectly as a result of very large course deviations, typically as a consequence of a sudden side-wind experienced when emerging from the shelter of a building.

Course deviation is a matter of dynamic rather than static instability. In most cases, the driver would be quite capable of holding the car on a steady line if the aerodynamic loads were applied slowly enough. Articulated trucks and car–caravan combinations are particularly prone to wind-induced dynamic instabilities, which can occur even in light winds. In this case, driving experience is a factor, which highlights the fact that the driver represents an important but somewhat unpredictable component in the vehicle stability system. Road vehicle cross-wind response is a complicated interaction between aerodynamics, suspension, steering geometry and human reaction.

FORCES AND MOMENTS AND THE AXES SYSTEMS FOR A ROAD VEHICLE

Thus far, lift and drag forces have been the main concern, but for cross-wind effects it is also necessary to take account of side force S, and the three moments, pitching M, rolling R and yawing Y. The complete system of moments and forces is illustrated in Fig. 10.1. A further complication, not previously mentioned, is that there are four sets of axes to deal with. The four possible longitudinal axes are shown in Fig. 10.2. Each of these axes has associated upward and sideways axes.

Drag is defined as the force resisting motion, and should preferably be taken as acting along the direction of travel axis. In stability analysis, however, it is often more convenient to use forces and moments resolved along the body axes, and in this book side force will be defined as the force normal to the body longitudinal axis, as shown in Fig. 10.1. Care has to be exercised when using published data, because there is a tendency to mix up the axis systems. Fortunately, under steady straight-line driving conditions, the direction of travel and car body axes coincide, in which case any ambiguity of definition becomes unimportant.

Yaw angle is normally defined as the angle between the car longitudinal axis and the relative wind direction, as shown in Figs. 10.1 and 10.3. The relative wind velocity is determined from the resultant of the wind velocity relative to the road, and the vehicle velocity relative to the road.

The centre of pressure defines the location of the line of action of an aerodynamic force (see Fig. 10.3). The centre of pressure of the resultant side force will not generally be at the same longitudinal location as the centre of pressure of the resultant lift force.

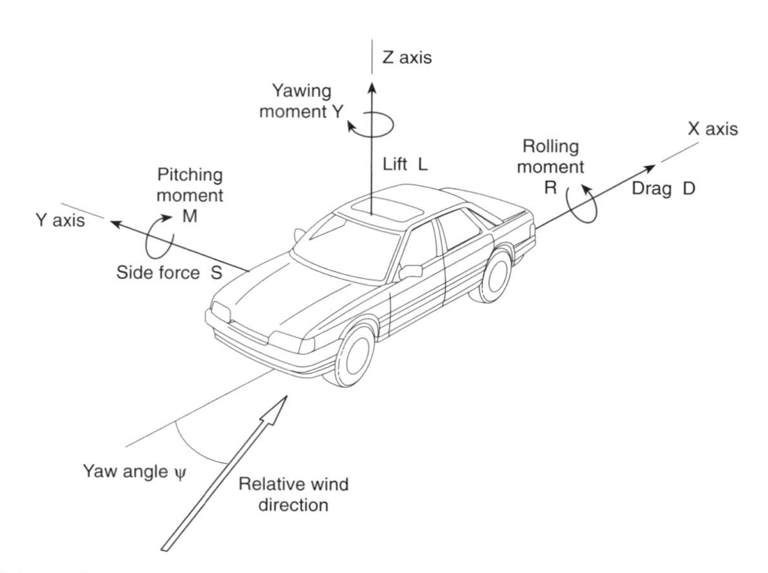

Fig. 10.1 Definition of forces and moments acting on a vehicle.

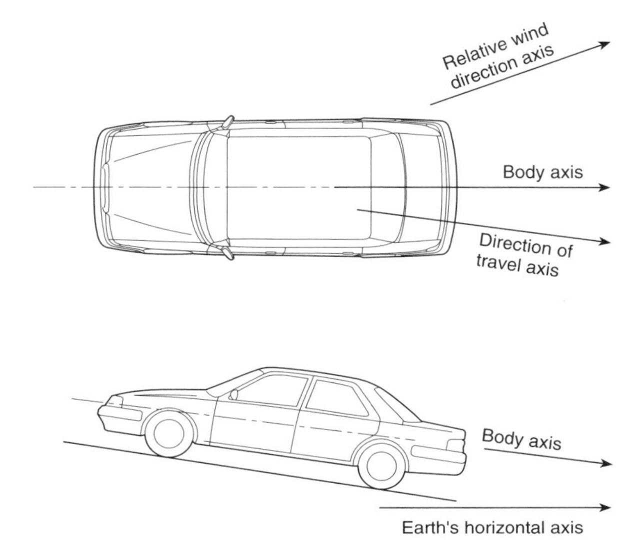

Fig. 10.2 The four sets of longitudinal axes.

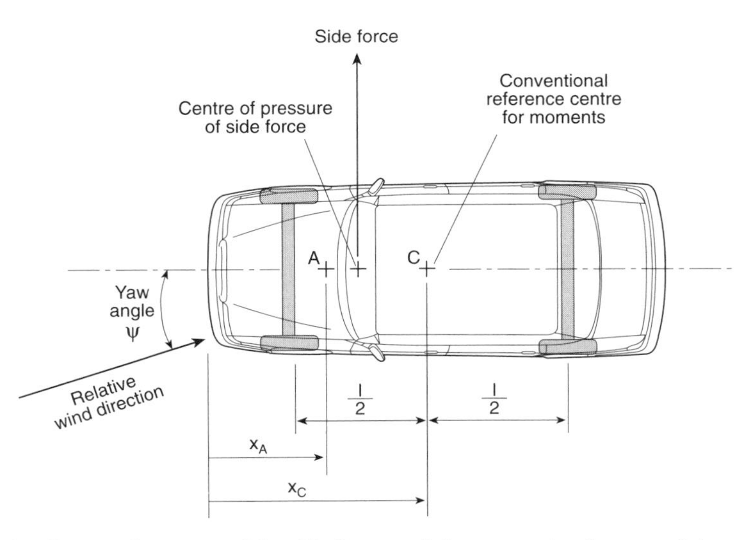

Fig. 10.3 Centre of pressure of the side force, and the conventional centre of the reference axes.

SIDE FORCE, YAWING MOMENT AND ROLLING MOMENT COEFFICIENTS

Just as it is convenient to use dimensionless lift and drag coefficients, it is also useful to define equivalent coefficients for side force and the pitching, rolling and yawing moments described above. Thus, the side force coefficient C_S is defined as

$$C_S = S/(\tfrac{1}{2}\rho V^2 A)$$

where S is the side force and the other symbols have their usual meaning. (Note that some authorities use the side area for this coefficient, but it is less confusing to use the frontal projected area for all the coefficients.)

The moment coefficients are given by:

Pitching moment coefficient $C_M = M/(\tfrac{1}{2}\rho V^2 A l)$

Yawing moment coefficient $C_Y = M/(\tfrac{1}{2}\rho V^2 A l)$

Rolling moment coefficient $C_R = M/(\tfrac{1}{2}\rho V^2 A l)$

Since moments have the dimensions of force × length, the moment coefficients have to be divided by an additional length l, for which the wheelbase dimension (distance between front and rear axles) is usually (but not always) used. The yawing, rolling and side force coefficients are referred to as the **lateral coefficients**.

A variety of reference areas and length dimensions are used by different authorities, but in this book the reference area A will always be taken as the projected frontal area, and the length l will be the wheelbase.

Note that the value of any moment depends upon the position of the axis about which the moment is measured. In this book the axes of the moments will be located at a 'reference centre' midway between the front and rear wheels, on the vehicle centre-line and at road level, as shown in Fig. 10.3. This is the most popular location, but care should be taken when referring to research papers, as authors sometimes take moments about the centre of gravity.

The sign convention for the forces is indicated in Fig. 10.1. A further source of some confusion is that it is also conventional to describe the yawing and rolling moments as positive if the vehicle tends to yaw or roll away from the wind direction.

Knowing the side force coefficient C_S, and the yawing moment coefficient about the reference centre $C_Y(C)$, it is a simple matter to calculate the yawing moment coefficient $C_Y(A)$ about any other axis position A by using the expression:

$$C_Y(A) = C_Y(C) - C_S(x_C - x_A)/l$$

The lengths x_C, x_A and l are defined in Fig. 10.3.

VARIATION OF COEFFICIENTS WITH YAW ANGLE

Bowman[5] studied a large number of car models in a wind-tunnel and made the following general conclusions.

1. The side force coefficient C_S varies linearly with yaw angle ψ up to at least 40°, then curves away to reach a maximum at around 60°.
2. The yawing moment coefficient C_Y roughly follows a relationship of the form $C_Y = k_y \sin 3\psi$, and thus reaches a maximum at around 30°.
3. the rolling moment C_R follows a curve roughly of the form

$$C_R = k_{r1} \sin \psi + k_{r2} \sin 3\psi$$

Bowman shows plots of the various coefficients against yaw angle for a large number of cars, but many of these relate to styles that are now obsolete. Figure 10.4 shows side force, yawing moment and rolling moment coefficients for a typical medium/large European car (based on data presented by Howell[13]). The measurements were made in the MIRA full-scale wind-tunnel.

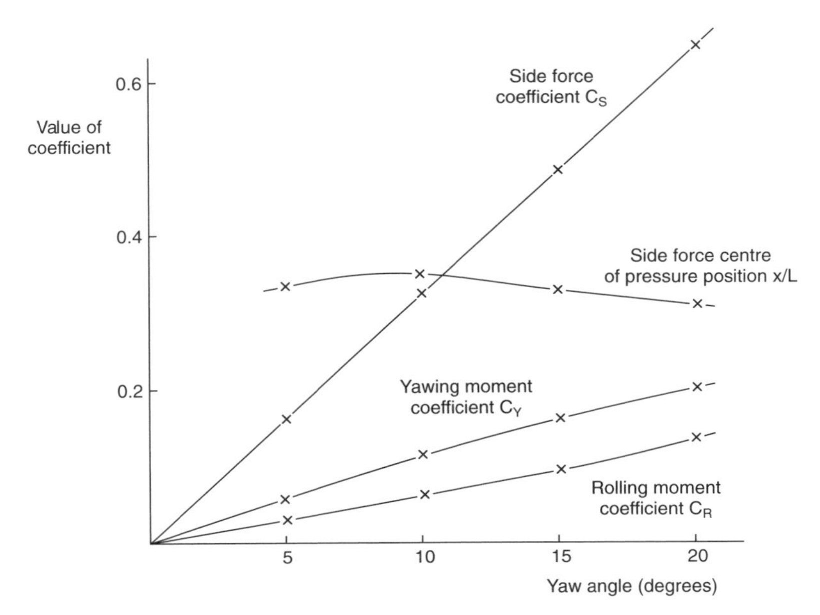

Fig. 10.4 Lateral coefficients for an early series Rover 800. (*After Howell*[13])

OVERTURNING DUE TO SIDE-WIND GUSTS

Aerodynamic rolling and overturning moments depend on the distribution of pressure around the vehicle. Figure 10.5 shows the local pressure distribution in a vertical plane passing through the roof leading edge of an early series Rover 800 (from Howell[13]). Results for two yaw angles and zero degrees are shown. The overall height of the car above the road was 1.39 m (4.56 ft), and it will be seen that the side force is concentrated towards the upper part of the body at this location, with the centre of pressure at about 1 m above the road. It should be remembered, however, that this cross-section is nearly the tallest part of the vehicle, so the centre of pressure of the overall side force is somewhat lower. The measurements used in Fig. 10.5 were made in a wind-tunnel, where the wind speed was uniform. In real road driving conditions, the wind speed varies with height, and this tends to move the side force centre of pressure up a little.

Most wind-related accidents involve some form of dynamic instability, but before considering that complex problem, it is worth looking at conditions necessary for a vehicle to simply blow over in the purely static or steady case. Overturning occurs when the vehicle rotates about the point of contact of the lee-side wheels. It can be seen from the simplified model of Fig. 10.6 that for a vehicle to overturn due to the direct action of a steady side-wind without any dynamic effects,

$$\tfrac{1}{2}\rho V_r^2 A[C_S z_{cp} + C_L(y_{cp} + t/2)] > Wt/2 \qquad [10.1]$$

where V_r is the resultant of the wind velocity and the driving velocity, and z_{cp} etc. are defined in Fig. 10.6. The effect of the lift component is normally quite small, but should not be ignored.

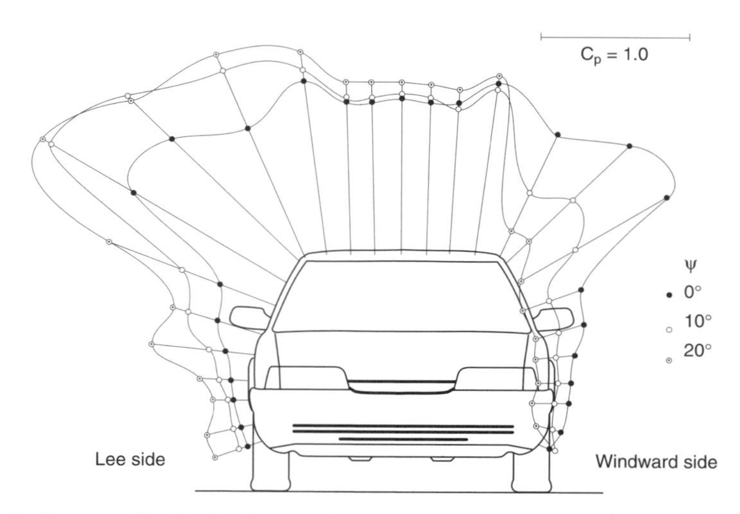

Fig. 10.5 Pressure distribution in a vertical plane passing through the roof leading edge for an early series Rover 800.
(*After Howell*[13])

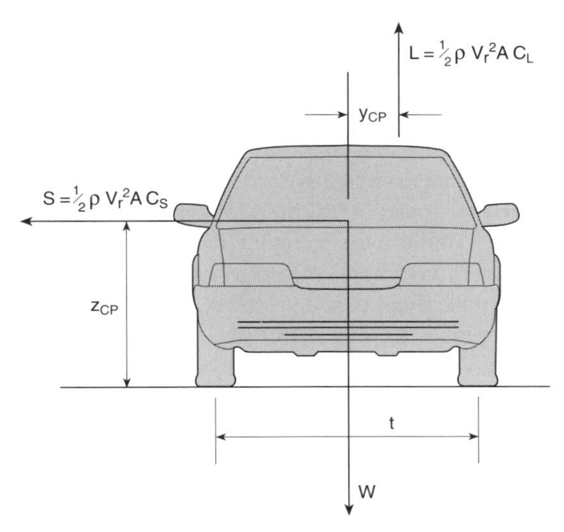

Fig. 10.6 Force required to overturn a vehicle in a purely static case. The vehicle will start to overturn when the moment of the side and lift forces about the leeward side wheels exceeds the moment due to the weight.

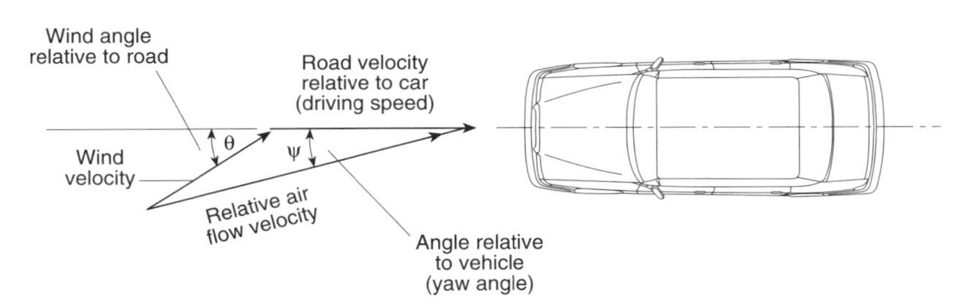

Fig. 10.7 The yaw angle depends on the wind speed and angle relative to the road, and the car driving speed. The lift and side force coefficients and the associated centres of pressure all vary with the yaw angle.

The side force and lift coefficient values in a cross wind depend on the yaw angle, which as may be seen in Fig. 10.7 depends on the wind speed, the driving speed, and the angle of the wind relative to the road. The values of y_{cp} and z_{cp} are also functions of yaw angle. An added complication is that the vehicle will heel over some way on its springs before the windward set of wheels starts to leave the road, and this change of geometry will also affect the aerodynamic moments. All these factors vary considerably according to body style, and the calculations for any vehicle require the use of wind-tunnel data relevant to that particular model. Once the dynamic effects and the driver reaction have been included, the problem of analysis becomes quite complex (see Baker[1]).

In their analysis of wind-related road accidents Baker and Reynolds[3] looked in detail at a number of accidents involving high-sided commercial vehicles. Estimated maximum wind gust speeds associated with these accidents ranged from 23 m/s to 33.7 m/s. It may be assumed that in most of these cases, dynamic effects played a significant part; however, using the condition expressed in equation [10.1], and taking typical published data for a lightly loaded high-sided truck, it can be computed that at a driving speed of 90 km/h (56 mph), overturning of this type of vehicle could occur even due to purely static effects in this range of wind speeds.

For vehicles with a very low centre of gravity, it is possible that the side force might be sufficient to cause them to slide sideways before being large enough to produce an overturning moment, but in this case, the wheels on one axle will slide before the other, so the vehicle will tend to yaw, and possibly go out of control. For this case to arise, the centre of gravity needs to be very low, as for example on a racing sports car.

ROAD VEHICLE LATERAL (SIDEWAYS) STABILITY

Road vehicle stability is yet another area where aeronautical engineers initially thought that it would be an easy matter to adapt their sophisticated methods of analysis to what appeared to be the relatively simple problem of motion in two dimensions rather than three. Once again, they were disappointed; to start off with, an aircraft is considered to be stable if, after a disturbance, it returns to its original attitude and direction of motion. For a road vehicle, these criteria alone would not be satisfactory, because to avoid danger of collision with other vehicles, it is also necessary to keep to a steady path. On motorways, the lateral accuracy required is in the order of 1 m (3 ft), but a tighter tolerance applies on most other roads.

YAWING RESPONSE TO SIDE-WINDS

On an aircraft, if the line of action of the side force produced by a gust lies behind the centre of gravity, the response is considered to be stable because the machine will tend to yaw towards the direction of the gust, and thus towards the relative flight direction. In a car, however, the response forces produced are not just aerodynamic, as there are also reaction forces from the road; the steering caster effect, for example, has a tendency to steer the vehicle away from the direction of the applied side force. Other aspects of the suspension geometry also influence the response of the vehicle. Furthermore, whilst turning towards the direction of a gust is considered a stable response in the case of an aircraft, a strong tendency to weathercock towards the wind does not represent a desirable condition in a road vehicle, which needs to maintain a constant path. The ideal yawing response would be no response; that is, the vehicle would simply continue on its path.

Fig. 10.8 Tail fins on a vintage Cadillac
Although only a popular styling gimmick at the time, they could well have improved the yawing stability. Is a comeback due?

For most types of conventional vehicle, the centre of pressure is located well forward, and the side force produces a tendency to yaw away from the wind direction. As explained in Chapter 3, to resist a side force, the tyres have to adopt a slip angle relative to the direction of motion, which implies that to maintain a steady path, the vehicle needs to be yawed slightly towards the wind gust direction; some corrective action is therefore normally required from the driver.

Moving the centre of pressure aft nearly always reduces the tendency of the vehicle to yaw away from the wind, and to achieve this, the side area at the rear needs to be increased. Perhaps tail fins (see Fig. 10.8), a popular styling gimmick of the 1950s, will one day make a comeback. Very large fins have sometimes been used on racing and record-breaking cars, but as mentioned previously, an excessive tendency to yaw towards the wind is undesirable, since the vehicle path and wind direction do not generally coincide.

YAWING MOMENT MAGNITUDE

There is one point on a vehicle where an applied side force would produce a neutral yawing response; that is, no tendency to yaw. The location of this point depends principally on the suspension and steering geometry and the position of the centre of gravity. It also varies with loading conditions and speed. As described above, however, aerodynamic yawing moments are usually measured about an axis located at the wheelbase centre (i.e. midway between the axles), so the numerical values thus obtained do not directly give an indication of stability. It is the moment about the point of neutral response that determines the strength and direction of the yawing response to a side load, not the moment about some arbitrarily chosen axis centre. Two vehicles with the same measured aerodynamic yawing moment about the conventional axis could have quite different yawing responses if their centre of gravity positions and suspension geometry were different.

SIDE LOAD DISTRIBUTION IN SIDE-WIND CONDITIONS

Overall side loads and yawing moments can be measured quite easily in a wind-tunnel, but is is also important to see what contributions are being made by different parts of the bodywork. Figure 10.9 shows the distribution of side force along the waistline of the early series Rover 800, obtained from pressure measurements. It will be seen that in the yawed condition (i.e. in the presence of a side-wind), there is a large concentration of side force near the front of the vehicle. Note, however, that the centre of pressure of the side force is not as far forward as might be suggested by this picture, because its location also depends on distribution of side area. Figure 10.4 given earlier relates to the same vehicle, and shows that the centre of pressure of the overall side force is around one third of the car length from the front. This location is fairly typical of a notchback car.

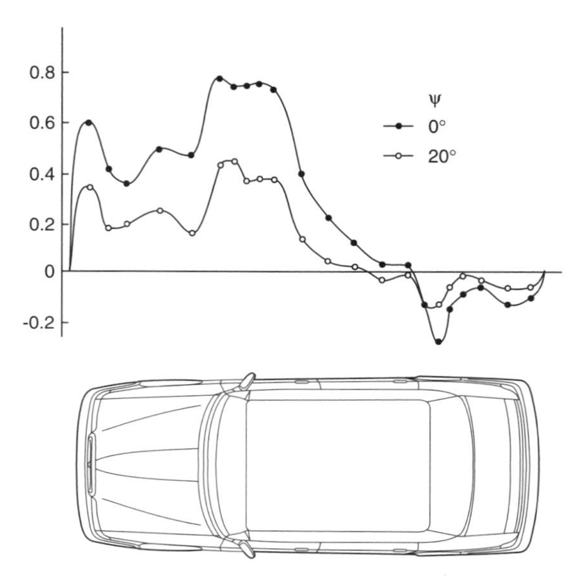

Fig. 10.9 The side force distribution around the waistline of an early Rover 800. (*After Howell*[13])

INFLUENCE OF BODY STYLE

Because the distribution of side area affects the position of the side force centre of pressure, the yawing response is sensitive to the body style. Squareback station wagon vehicles with the area concentrated to the rear produce lower yawing moments (about the reference centre) than notchback saloons. Gilhaus and Renn[11] tested variants of a wind-tunnel model in the three most common car body forms. The notchback (saloon) version gave a 12 per cent greater yawing moment than the squareback station wagon at any yaw angle, with the hatchback giving a moment value about midway between that of the other two forms. Note, however, that the figure of 12 per cent does not represent a qualitative indication of the relative yawing stability, because the axis for the yawing moment measurements was in the conventional mid-wheelbase position.

Single-box vehicles such as buses and vans with well-rounded leading corners tend to behave like an aerofoil at low angles of yaw, and the centre of pressure of the side force is located at around one quarter of the vehicle length from the front. This again means that the centre of pressure is well forward, making such vehicles relatively less stable than two- or three-box forms.

As the yaw angle increases, the flow eventually separates on the lee side, and the vehicle will then act more like a stalled aerofoil, with the centre of pressure moving aft, thereby reducing the yawing response to gusts. For this reason, there is some advantage in having front edges that are sufficiently rounded to produce low drag at zero yaw angle, but sharp enough to ensure separation at fairly small yaw angles. The problem with this solution, however, is that any sudden change in the centre of pressure position and side force caused by the separation can produce an unpredictable 'twitchiness' that can upset the driver's response, thereby leading to a dynamic instability of control. It is therefore important to try to ensure that the separation increases progressively with yaw rather than suddenly, and that it occurs at small yaw angles, so that a large side force does not develop before separation. Considerable design ingenuity and extensive wind-tunnel development is required.

Gilhaus and Renn[11] made a detailed study of the influence of many styling aspects on the lateral coefficients. Significant reductions in yawing moment were obtained by:

- lowering and squaring off the lower front valance
- rounding the tops of the front wings (fenders)
- rounding the A-posts

Rounding the rear corners increased the yawing moment, which is a pity, because such rounding has a beneficial effect on drag. In general rounding of the front end does not adversely affect the lateral stability, and it is the rear end that can be a source of problems.

Rolling moment was found to be reduced by:

- rounding of corners, particularly on lengthwise edges
- increasing the tumblehome (the front view taper of the upper part of the vehicle)

In addition to describing the effects of styling features qualitatively, attempts have been made to produce quantitative predictions of values of the various lateral coefficients, C_Y etc. Bowman[5] suggested ranges of values for the constants in his expressions (see p. 215) for the variation of the lateral coefficients with yaw angle, relating these to characteristic body styles. Unfortunately, as most of these body styles are now outdated, many of the data are no longer relevant.

Following the success of White at MIRA in producing a drag coefficient prediction method based on rating design features, as described in Chapter 4, attempts have been made by Carr *et al.*[7] to apply a similar approach for the prediction of yawing and side force coefficients. Initial results were, however, less successful than for drag and lift coefficients.

Theoretical methods for the prediction of the lateral coefficients have also been developed (Hucho and Emmelmann[14]; Tran[18]). Considerable simplifications were involved, and the models cannot account well for the effects of styling details. An empirical approach based on experimental data and an understanding of the flow features seems more promising.

One problem with any prediction method is that since all aerodynamically styled cars tend to end up with similar overall shapes, the accuracy of the method needs to be high to be of any real value. The simplest approach is to use values from any existing vehicle that closely matches the shape of the proposed new design.

THE INFLUENCE OF AERODYNAMIC LIFT ON STABILITY

Until the 1980s, most road-going vehicles produced positive upward lift. The lift coefficient of domestic cars was often greater than the drag coefficient, with C_L values typically ranging from 0.3 to 0.7. Reducing the lift or even producing a down force is found to improve the cross-wind stability (see Flegl and Rauser[9] and Buchheim *et al.*[6]). It seems logical that if the vehicle is firmly in contact with the road, it will be more stable than if part-way to take-off. It is also found that the pitching moment should preferably be positive (nose up), because this tends to give a safe oversteer characteristic. A positive pitching moment implies less lift (or more down force) on the rear axle than the front.

In addition to the reduction in drag that usually accompanies a reduction in lift, the favourable stability characteristics of low or negative lift provide good reasons for reducing the (positive upward) lift coefficient of vehicles. The relevant techniques for domestic cars are described in Chapter 4, and for racing cars in Chapter 6.

RATES OF CHANGE OF SIDE FORCE AND MOMENTS - AERODYNAMIC DERIVATIVES

An important factor in lateral stability analysis is the rate at which the forces and moments vary with yaw angle, and it is necessary to measure quantities such as $\mathrm{d}C_S/\mathrm{d}\psi$; these are known as aerodynamic derivatives. A large value of the derivative means that the force or moment changes rapidly with angle, so the vehicle will be very sensitive to yaw angle changes. In aircraft stability analysis, the derivative values are almost constant for all normal flight conditions, which encompass only a small range of operating angles, and this greatly simplifies the stability analysis. With road vehicles, however, wind gust speeds can be similar to the driving speed, and the gust can approach from almost any angle. Over such a wide range of yaw angle, the derivatives will not have a constant value, and the problem of stability analysis is much more complex than that for simple aircraft.

A further difficulty is that aerodynamic derivatives are known to be sensitive to rate of movement; that is, the value of the force or moment depends not only on the angle of yaw, but on the rate of yaw. Wind-tunnel measurements by Garry and Cooper[10] show that aerodynamic forces are sensitive to rate of yaw, even at very slow yaw rates. Measurements were made of the forces and moments on a model mounted on a slowly moving turntable in a wind-tunnel. The moving-model values were found to be similar to those measured statically, but in general, lagged behind the static measurements, as shown in Fig. 10.10. This was true even at turntable rotation rates as low as 0.25 degrees/second. Clearly, the air flow takes some time to fully establish itself to a new pattern. Bearman and Mullarky[4] measured forces and moments on a model in an oscillating flow generated by sinusoidally moving vanes placed upstream in a wind-tunnel. The measured oscillating forces were compared with those calculated from values obtained statically. At high oscillation

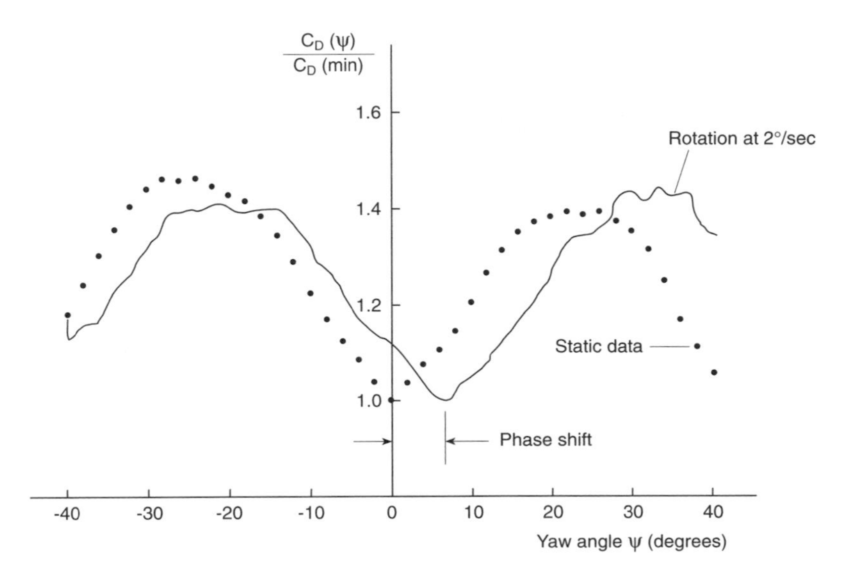

Fig. 10.10 Time lag in forces measured on a slowly rotating model compared with data obtained by static (steady) measurements. Drag data from the model moving slowly in yaw are similar to the data measured under static conditions, but shifted in phase. (*After Garry and Cooper*[10])

rates, it might be expected that the forces in the oscillating flow would be lower than the statically determined ones, as the gust size becomes smaller than the vehicle, but as with Garry and Cooper's work, the values in the oscillating flow were found to be lower than those predicted directly from static measurements even at the lowest oscillation rates.

The fact that the aerodynamic derivatives are rate-sensitive complicates the stability analysis and poses problems of experimental measurement. Fortunately, the dynamically measured values are normally found to be lower than those obtained from static tests, so calculations based on static data are conservative; that is, pessimistic.

EXPERIMENTAL METHODS OF ASSESSING CROSS-WIND DYNAMIC EFFECTS

Because of the limitations of static data, some efforts have been made to use moving models in wind-tunnels. One suitable experimental arrangement has been installed at the University of Nottingham (Humphreys and Baker[15]). The model is placed on a trolley on a track, and is fired across the tunnel by a catapult arrangement. A miniature six-component balance and data acquisition equipment are mounted on the trolley. Figure 10.11 shows a model of an articulated truck during its passage across the tunnel. A simulated atmospheric boundary layer can be used, and the technique overcomes the fact that it is otherwise impossible to simulate the skewing of the wind velocity described previously. Experiments by Humphreys and Baker[15]

Fig. 10.11 Moving-model arrangement to assess dynamic cross-wind effects in the wind-tunnel at Nottingham University. The model is attached through a slot in the floor to a track underneath, and is catapulted across the tunnel, emerging from the small hole in the left of the picture. The gravel in the foreground is used to help the generation of realistic scaled wind turbulence.
(*Photograph courtesy of Dr C. Baker, University of Nottingham*)

indicate that dynamically obtained data do not differ radically from those obtained statically once a steady condition is reached, although there is a delay in reaching that steady condition. It was found for example that when a vehicle entered suddenly into a cross-wind, the static value of side force was reached within two to three vehicle lengths. The dynamic tests can of course reveal a great deal of information that cannot be obtained from static tests, such as response to transient effects, and the influence of terrain variability.

Full-scale vehicle test tracks such as that of the motor industry research association (MIRA) have facilities for producing cross-winds artificially. The MIRA arrangement consists of an array of exhaust jets from a gas turbine directed to blow across the test track. The vehicle is driven past the cross-flow, and course deviations are recorded.

THE DRIVER AS AN ELEMENT IN THE CONTROL LOOP

Figure 10.12 shows typical driver-controlled responses to a sudden side gust compared with the uncontrolled natural responses of the vehicle for the case of the steering wheel left free. After a short delay due to the driver's reaction time, which

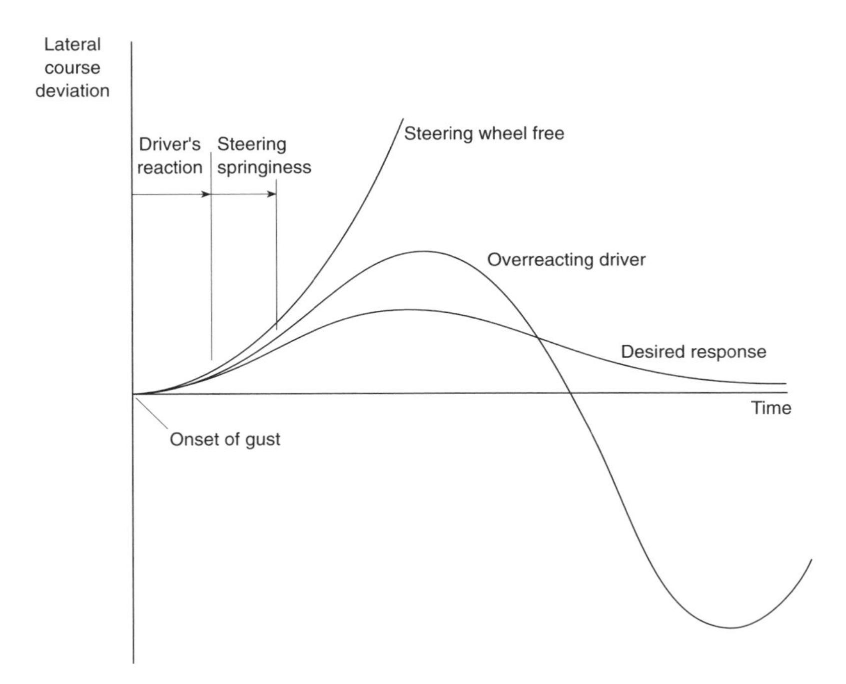

Fig. 10.12 Driver-controlled and stable reaction and dynamically unstable overreaction, compared to natural unstable mechanical responses of a vehicle to a sudden side gust. (*After Emmelmann*[8])

can vary greatly between individuals, a steering input begins to be applied. A further delay occurs as the springiness in the mechanical system is taken up. The course deviation rate then starts to decline. In the desired response, the vehicle then moves back on to its original line. The time or distance taken to return to the line and the deviation are indicators both of the vehicle's stability characteristics and of the driver's skill. An unacceptable response is also shown in Fig. 10.12. Here, the driver is overcorrecting, and producing an oscillation of increasing amplitude. There is therefore always an element of subjectivity in cross-wind testing on the track, and ideally several test drivers are required, each driving a range of vehicles for comparison.

The response of the vehicle is critically dependent on the driver, who always forms an active element in the control loop, unlike a pilot, who only needs to respond to large inputs and make occasional small corrections. Even the driving position can influence the stability of the vehicle/driver system. In a forward-control vehicle, the driver will be subjected to much greater lateral accelerations than if he were sitting near the centre of gravity, and may therefore react more quickly.

From the above, we can see that we cannot effectively separate the aerodynamic stability effects from the characteristics of the suspension, steering and driver reaction. A full mathematical analysis of road vehicle dynamics is thus extremely complex (see Baker[1]), and should take account of variables such as the tyre adhesion properties, which are sensitive to a wide range of parameters, including road surface and weather conditions. The driver's response is subject to a great deal of uncertainty and variability.

LATERAL EFFECTS OF ADJACENT AND OVERTAKING VEHICLES

When one vehicle overtakes another, they mutually influence the flow field around each other. Thus even when there is no cross-wind, the asymmetry of the flow can cause side forces and yawing moments. Figure 10.13 shows the side force and yawing moments on a car when overtaking a large truck. These data are taken from wind-tunnel measurements by Howell[12]. The models were moved relative to each other step by step, so no dynamic effects were simulated.

It will be seen from Fig. 10.13 that as the car comes up towards the large truck it is initially pulled towards it, but as the car draws alongside the truck, the force rapidly changes sign and the car is repelled. The yawing moment variation follows a similar pattern. The sudden change of direction and increase in magnitude as the car passes the truck make it difficult for the driver to maintain a steady line. A more recent study of the problem of vehicles overtaking in a tunnel may be found in Minato and Ryu.[17]

Erratic changes in the forces and moments also occur when a small vehicle manoeuvres either in front of or behind a larger one as shown in Fig. 10.14. The data for this figure are taken from wind-tunnel experiments conducted by the author and M. Naji (1981); the experimental arrangement used was similar to that of Howell, except that simple rectangular blocks were used instead of vehicle models.

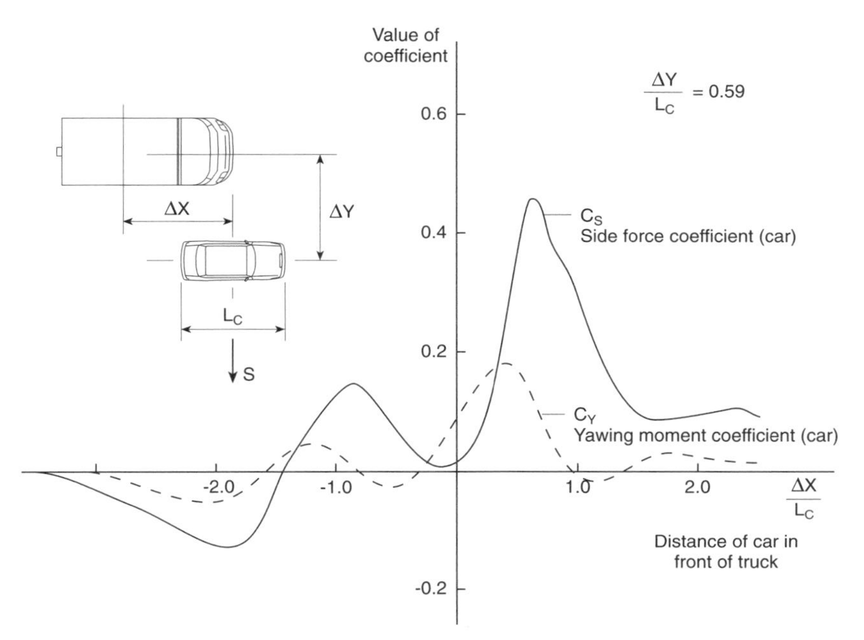

Fig. 10.13 Side force and yawing moment changes occurring when a small vehicle overtakes a larger one.
(Based on wind-tunnel data in Howell[12])

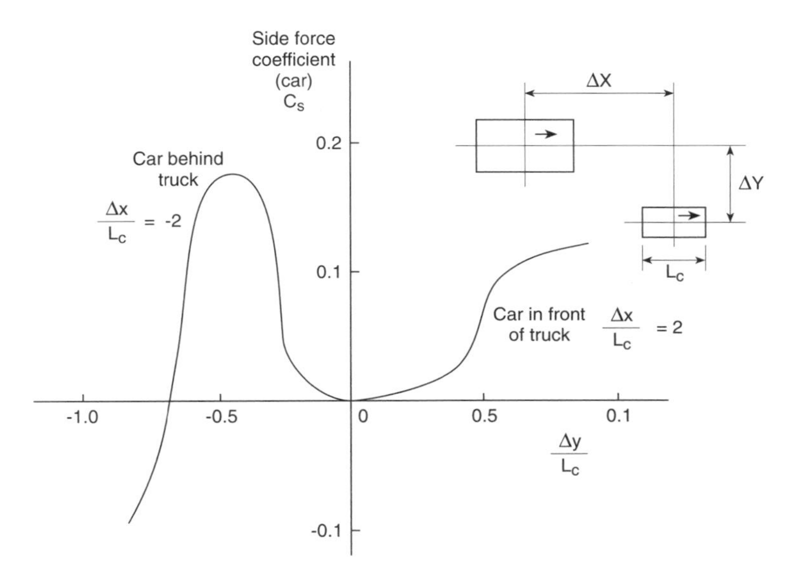

Fig. 10.14 Side force and yawing moment changes occurring when a small vehicle manoeuvres in front of or behind a larger one.
(Based on data from wind-tunnel measurements on simple block shapes by the author and M. Naji)

SOME SIMPLE RULES FOR GOOD AERODYNAMIC STABILITY

It is not possible to reduce the stability equations to a meaningful simple form, but some aerodynamic factors that influence the stability can be isolated, and some general rules for good aerodynamic design can be identified.

The first rule is that the centre of pressure of the side force should be well aft. In practice, most vehicles have the centre at about one quarter to one third of the car length from the front. Increasing the side area at the rear can help to move the centre of pressure aft. The use of front-wheel drive encourages a forward location of the centre of gravity, which also helps the yawing stability.

It is important to recognize the importance of the driver in the stability loop; he must have some indication of the presence of gusts, and the need to take corrective action. The side force response should preferably be a linear one; sudden changes in the side force and yawing moment produce a twitchiness that is difficult to cope with. Such changes usually occur as a result of changes in the position of flow separation lines. An unstable location of the separation line at the rear of the vehicles is a common source of cross-wind instability, and it may be necessary to use sharp corners to maintain a fixed or predictable separation line. Figure 10.15 shows the raised rear side-window trim element on a Ford Sierra, which was introduced on later models to help improve the lateral dynamic stability by fixing the separation line.

Most forms of vehicle can suffer from sudden lateral flow separations leading to poor handling, and this has led to the realization that experimental examination of the cross-wind behaviour is just as important as drag reduction. It is also even more important that the wind-tunnel testing is undertaken at a relatively early stage in the design process, because it is not easy to provide 'fixes' for poor lateral characteristics.

Fig. 10.15 Raised trim elements on the rear side windows of later models of the Sierra helped to fix the separation position, and improved the cross-wind stability characteristics. The small wing also aids stability by reducing rear lift.

REFERENCES

1. Baker C. J., Ground vehicles in high cross winds, parts 1, 2 and 3, *Journal of Fluids and Structures*, 5, pp. 69–111, 221–41, 1991.
2. Baker C. J., The behaviour of road vehicles in unsteady cross winds, *Journal of Wind Engineering and Industrial Aerodynamics*, Vol. 49, 1993, pp. 439–48.
3. Baker C. J. and Reynolds S., Wind-induced accidents of road vehicles, *Accident Analysis and Prevention*, Vol. 24, No. 6, pp. 559–75, 1992.
4. Bearman P. W. and Mullarky S. P., Aerodynamic forces and moments on road vehicles due to steady side winds and gusts, *Proc. Vehicle Aerodynamics Conf.*, RAeS, Loughborough, 18–19 July 1994.
5. Bowman W. D., Generalizations on the aerodynamic characteristics of sedan type automobile bodies, SAE paper No. 660389.
6. Buchheim R., Maretzke J. and Piatek R., The control of aerodynamic parameters influencing vehicle aerodynamics, SAE paper No. 850279, 1985.
7. Carr G. W., Atkin P. D. and Somerville J. (MIRA), An empirically-based prediction method for car aerodynamic lift and side forces, *Proc. Vehicle Aerodynamics Conference*, RAeS, Loughborough, July 1994, pp. 33.1–16.
8. Emmelmann, H. J., Driving stability in side winds: in *Aerodynamics of Road Vehicles*, ed. W. H. Hucho, Butterworth, London, 1987, pp. 216–35.
9. Flegl H., and Rauser M., High performance vehicles, in *Aerodynamics of Road Vehicles*, ed. Hucho W. H., Butterworth, London, 1987.
10. Garry K. P. and Cooper K. R., Comparison of quasi static and dynamic wind-tunnel measurements on simplified tractor–trailer models, *Proc. 6th Colloquium on Industrial Aerodynamics*, Aachen, 19–21 June, 1985.
11. Gilhaus A. M. and Renn V., Drag and driving stability related aerodynamic forces and their interdependence – results of measurements on 3/8 scale basic car shapes, SAE paper No. 860211, 1986.
12. Howell J. P., The influence of the proximity of a large vehicle on the aerodynamic characteristics of a typical car, in *Advances in Road Vehicle Aerodynamics*, ed. H. S. Stevens, BHRA Fluid Engineering, 1973, pp. 207–21.
13. Howell J. P., The side load distribution on a Rover 800 saloon car under crosswind conditions, *Proc. Wind Engineering Society Conference*, Warwick, 1994.
14. Hucho W. H. and Emmelmann H. J., Theoretical prediction of the aerodynamic derivatives of a vehicle in cross-wind gusts, SAE paper No. 730232.
15. Humphreys N. D. and Baker C. J., The aerodynamic forces and moments on lorries in cross winds, Extended abstract, *Inaugural Conference of the Wind Engineering Society*, Downing College, September 1992.
16. Kobayashi T. and Kitoh K., Cross-wind effects and the dynamics of light cars, in *Impact of Aerodynamics on Vehicle Design*, ed. M. A. Dorgham, *Int. Journal of Vehicle Design*, Special Publication SP3, 1983, pp. 142–57.
17. Minato K. and Ryu H., Aerodynamics of vehicles in tunnels – flow visualisation using the laser light sheet method and its digital image processing, SAE paper No. 910314, 1991, 00123–131.
18. Tran V. T., A calculation method for estimating the transient wind force and moment acting on a vehicle, SAE paper No. 910315, 1991.

CHAPTER 11

WIND-TUNNEL AND ROAD TESTING

There are no simple theoretical methods for predicting the aerodynamic characteristics of a road vehicle, and as described in the next chapter, computation methods still have considerable limitations; therefore, it is necessary to use wind-tunnel and road testing extensively.

WIND-TUNNEL TYPES

Wind-tunnels were originally developed primarily for aeronautical use, although in recent years, specialized tunnels intended exclusively for automotive work have been evolved, and these have a number of features that distinguish them from aeronautical tunnels.

Two different overall configuration of wind-tunnel exist. In the closed-return form, as typified by General Motors' Detroit tunnel shown in Fig. 11.1, air circulates round a continuous closed circuit. The alternative open-return arrangement used for the MIRA full-scale tunnel shown in Fig. 11.2 consists essentially of a tube open at both ends. Air is drawn from and returned to the surroundings.

The advantage of the open-return type is that it is relatively compact, and normally makes use of the open space of the enclosing building as its return circuit. Another advantage is that because the air is drawn from and discharged into a relatively large space, it does not tend to heat up. In the case of a closed-return tunnel, the kinetic heating produced by friction on the walls can cause a large temperature rise, which will adversely affect the stability of measuring instrumentation; note the cooler shown in Fig. 11.1. The open-return tunnel does however have two disadvantages; firstly, as the air is simply exhausted to the surroundings, its kinetic energy is wasted, and therefore more power is required than for a closed-return configuration. A second problem is that since the air starts from rest in the surrounding room at atmospheric pressure, the pressure in the working section must be lower than atmospheric (following the Bernoulli relationship). The working section interior must therefore be well sealed from the surroundings.

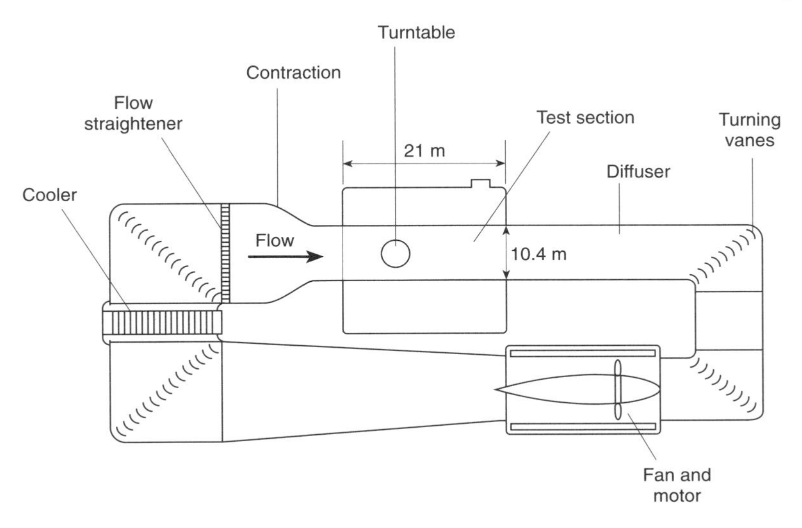

Fig. 11.1 A closed-return type of tunnel
The GM full-size vehicle tunnel with a working section 5.5 m by 10.4 m (18 ft by 34 ft).

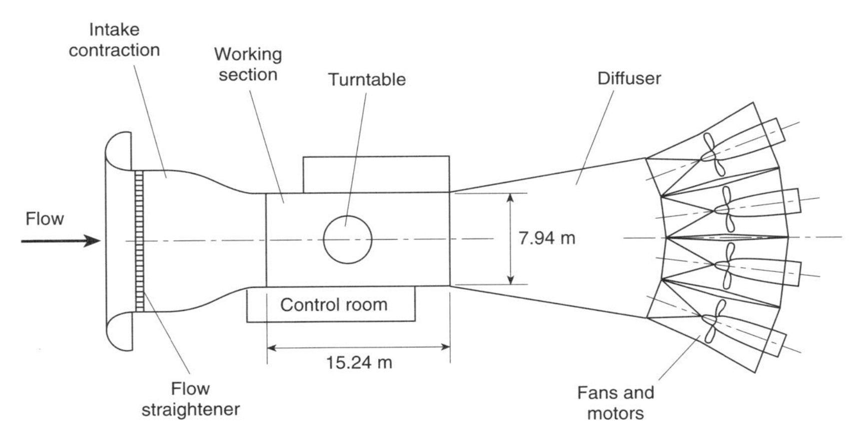

Fig. 11.2 An open-return type of tunnel
The MIRA full-size vehicle tunnel with a working section 4.42 m by 7.94 m (14 ft by 26 ft).

The two types of tunnel are capable of giving equally good results, but for practical reasons, if money is no limitation, the closed-return type is now usually preferred, and almost all of the larger manufacturers have installed this type in recent years. Apart from the return circuit, there are major differences in the form of the working section used, but these will be explained at the appropriate point in the text.

WIND-TUNNEL FEATURES

It may be seen from Figs. 11.1 and 11.2 that both types of tunnel contain a short contraction, where the air is speeded up, followed by a working section and then a diffuser, where the flow is slowed down again. The contraction ratio is defined as the ratio of two areas: that at the inlet to the contraction section and that at its outlet. In general, having a large inlet contraction ratio improves the steadiness and uniformity of the flow in the test section. In the case of a closed-return tunnel, however, a large contraction ratio requires the use of a large diameter return circuit. Turning vanes and screens help to reduce turbulence levels. The fact that the turbulence level is almost always higher in a tunnel than on the road (on a calm day) is one source of discrepancies between wind-tunnel and road test results.

Wind-tunnels are characterized by the cross-sectional dimensions of their working section. The GM tunnel in Fig. 11.1 has a massive working section 5.5 m high by 10.4 m wide (18 ft by 34 ft), which enables a full-size truck to be tested. The MIRA tunnel depicted in Fig. 11.2 is a slightly more modest 4.5 m by 8 m (15 ft by 26 ft), though still large enough to take most types of vehicle in full scale.

SOURCES OF ERRORS IN WIND-TUNNEL TESTS

It might be thought that by now, with all the vast experience of aeronautical work, wind-tunnel testing would produce totally reliable results. This is, however, not the case, and as will be seen later, well-established laboratories can produce significantly different results even when using the same model. There are three primary sources of error in wind tunnel-testing:

1. scale or Reynolds number effect
2. errors due to the fact that the road moves relative to the car, whereas the floor of the tunnel is normally stationary
3. errors due to blockage

Other important sources of error are

4. failure to model fine detail accurately
5. failure to model the effects of the flow through the cooling and ventilating systems
6. difficulties in measuring forces when the wheels are in contact with the road.

SCALE OR REYNOLDS NUMBER EFFECT

Figure 11.3 shows two thin almost flat plates placed parallel with a stream of air. One may be thought of as representing part of a full-size vehicle, and the other the corresponding part of a model. Transition from a laminar to a turbulent boundary layer will occur at the same distance from the leading edge on both plates, as illustrated. From the diagram, it will be seen that the scale model will therefore

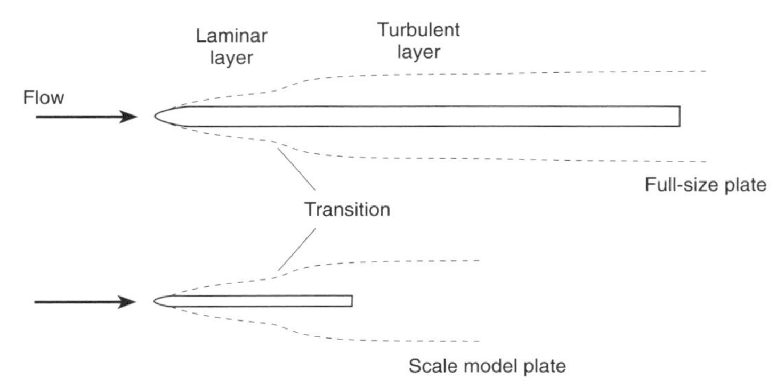

Fig. 11.3 Boundary layer and scale effect
Transition occurs at the same distance from the leading edge for both model and full-scale plate, so the flow features of the model are not a scaled representation of the full-size situation.

have a greater proportion of laminar boundary layer, and consequently a lower drag per unit of surface area than the larger one. The drag per unit area measured on the model is thus not representative of full scale.

To correct for the effect of scale, the model could be put in a stream of air moving faster than that for the full-size plate. This would increase the Reynolds number ($\rho V l/\mu$) and move the position of transition forward. If the speed were sufficiently high, the transition point could be moved forward to a scale position corresponding to that on the full-scale plate.

The same principle applies to all shapes, and it is found that to obtain similar flow patterns and features for both full-scale and model, the Reynolds number must be the same for both.

$$\frac{\rho V l}{\mu} \text{ model} = \frac{\rho V l}{\mu} \text{ full-scale}$$

Since the length l of the model is smaller than that of the full-scale vehicle, $\rho V/\mu$ must be made larger for the model to compensate.

Unfortunately, it may be seen that unless something is done about the air density or viscosity values, a 1/10 scale model would need to run at 10 times the full-scale speed. This, however, means that to simulate 160 km/h (100 mph), it would be necessary to run the tunnel at 1600 km/h (1000 mph), which is well above the speed of sound at sea level! Clearly the resulting supersonic conditions would ensure that the flow around the model would be nothing like that for the full-size car. In practice there are few automotive tunnels capable of speeds in excess of about 250 km/h (70 m/s or 156 mph); high-speed tunnels are expensive both to construct and to run.

TACKLING THE REYNOLDS NUMBER PROBLEM

For aeronautical work, as an alternative to running the tunnel at high speed, the Reynolds number can be raised by pressurizing the tunnel to increase the density, but the tunnel than becomes a large and extremely expensive pressure vessel. Very few such tunnels exist despite the generous funding that was available for aerospace defence work during the Cold War. As an alternative, the viscosity μ can be altered by cryogenic cooling, but this is an even more expensive solution.

For road vehicle testing, when accurate results are required, it is much cheaper to build an unpressurized tunnel large enough to take a full-size vehicle. The cost of building and running a tunnel large enough to accommodate a full-size car is not excessive for industrial users, and accurate wind-tunnel scale models can be just as expensive to construct as a real full-size car. Most major manufacturers now either have a full-size tunnel or have access to one. Table 11.1 lists some of the current large full-size tunnels and their salient features.

Table 11.1 Salient features of some larger full-scale automotive wind-tunnels; worldwide, there are around two dozen of such tunnels. A wide variety of forms is used.

Tunnel	Return circuit	Test section	Section dimensions	Max. speed (km/h)
NRC Canada	Closed	Closed	$83.6\,m^2$	200
GM USA	Closed	Closed	$5.5 \times 10.4\,m$	240
DNW Netherlands	Closed	Closed	$6 \times 8\,m$	220
Volkswagen Germany	Closed	Open	$4.9 \times 7.6\,m$	180
Lockheed Georgia USA	Closed	Closed	$35.1\,m^2$	406
MIRA UK	Open	Closed	$4.42 \times 7.94\,m$	133
Daimler Benz Germany	Closed	Open	$4.87 \times 7.38\,m$	270
Fiat Italy	Closed	Open	$30\,m^2$	200
Ford (Cologne) Germany	Closed	Open	$3.9 \times 6.1\,m$	182
Nissan Japan	Open	Closed	$21\,m^2$	119
BMW Germany	Closed	Slotted wall	$20\,m^2$	160
St Cyr France	Open	Slotted wall	$15\,m^2$	144
JARI Japan	Open	Closed	$12\,m^2$	205

SMALL WIND-TUNNELS

For fundamental studies and exploratory test programmes it is not necessary to achieve a very high absolute accuracy. In many cases, the purpose of the testing is essentially comparative; the manufacturer is merely interested in whether a modification reduces or increases the drag coefficient. For such tests, perfectly adequate results can be obtained using scale models in small wind-tunnels. The effect of using the wrong Reynolds number is not usually very large, as long as the test Reynolds number is above a certain critical value. As described in Chapter 2 (Fig. 2.6), for a circular rod there is a sudden drop in drag coefficient at a Reynolds number of around 5×10^5, after which the coefficient stays fairly constant. Many shapes show a similar effect at around the same Reynolds number (based on cross-stream depth rather than length), and for a car with a depth of 1.4 m (4.6 ft), the critical Reynolds number value corresponds to about a 1/5 scale model at 25 m/s (82 ft/s). It is, however, preferable to leave a margin, and 40 m/s (130 ft/s) at 1/4 scale is a popular standard.

Car shapes rarely show a dramatic change in C_D with Reynolds number, particularly when they have been optimized to produce mostly attached flow, and acceptable results can often be obtained with quite small models at low speeds. The variation of C_D with Reynolds number for a 1/8 scale model of a Group C sports racing car tested by the author[1] was shown in Chapter 2 (Fig. 2.7). In this figure, some variation in the drag coefficient value at low Reynolds numbers is evident, but at high tunnel speeds, the coefficient settles down to near the full-scale value. This behaviour is fairly typical of well-streamlined vehicles, but as described in Chapter 5, the drag coefficient of flat-fronted trucks and buses with rounded corners is very sensitive to Reynolds number. It is in addition difficult to achieve a high Reynolds number in truck testing, because the vehicles are relatively large, and thus the models have to be at a very small scale if they are to be fitted into a small wind-tunnel. Even most full-scale tunnels are too small to accommodate a large truck without causing blockage problems (described later). For this reason, trucks are often tested using 1/4 or even 1/2 scale models in full-size car wind-tunnels.

Small wind-tunnels can be extremely useful for fundamental studies, and most large laboratories have a tunnel suitable for 1/4 or 1/3 scale models to complement their full-scale tunnel. Because the tunnels cannot generally be run fast enough to produce the full-scale Reynolds number, they are usually run at almost maximum speed. High tunnel speeds produce large forces and pressure differences, which make measurements easier. The limiting factors in tunnel speed are noise, running costs and overheating.

A typical question asked by people unfamiliar with the subject is 'what speed should the tunnel be run at to simulate say 100 mph on the road?' From the preceding, it may be seen that for a 1/5 scale model, the true scale speed would be 500 mph, so as explained above, the answer has to be 'as fast as practically possible'.

TRANSITION STRIPPING

The main cause of discrepancies between full-size and small-scale model tests is that in the former case, much of the boundary layer will be turbulent, whilst in the latter, most of it will be laminar due to the lower Reynolds number. This influences both the surface friction drag, and the position of separation and reattachment. The deficiency can be partially corrected by attaching small strips of sandpaper to the surface of the model at a position where transition is predicted to occur in full scale. Since transition is normally very near the front, and the effect of incorrect positioning is usually quite small, the location of the strips does not have to be precise. Transition strips may be seen on the truck model shown in Fig. 5.7.

THE MOVING ROAD PROBLEM

Under real road conditions in still air, there is no relative motion between the road and the air, and thus no boundary layer is developed on the road. In a wind tunnel, however, a boundary layer is developed on the floor, and the model will be partially immersed in it, significantly affecting the results, unless some form of corrective action is taken. The methods normally employed are:

- use of a ground board – Fig. 11.4
- use of a belt moving at the same speed as the wind, with boundary layer removal upstream – Figs. 11.5 and 11.6
- use of a porous floor with suction – Fig. 11.7

THE GROUND BOARD METHOD

The ground board is the simplest method, and can give quite satisfactory results. The board is mounted well clear of the tunnel boundary layer, and although there is relative motion between the air and the board, the resulting boundary layer on the board is thin. To compensate for even this thin layer, the model can be raised by an amount corresponding to the boundary layer displacement thickness: the amount by which the boundary layer displaces the streamlines upwards.

It is important to ensure that differences in blockage of the flow above and below the board do not lead to a circulatory tendency around the board (air flowing at a different rate over the upper and lower surfaces). This tendency can be corrected either by equalizing the blockage amounts, or by means of a small flap on the trailing edge of the plate (see Fig. 11.4).

Garry[17] describes an investigation of the influence of plate length on drag coefficient measurements. His studies show that the coefficients are quite sensitive to the length of the plate downstream of the rear of the model. A length of around three times the model width appears to be the minimum necessary. The maximum

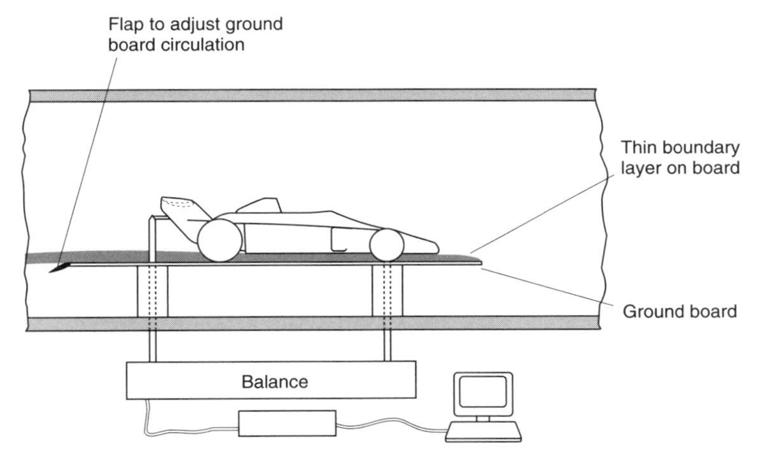

Fig. 11.4 The ground board method
The model is mounted above a board which spans the tunnel. This is the simplest method of reducing the ground boundary layer effect.

error occurs when the downstream length is roughly equal to the model width. The importance of the downstream ground board length is presumably due to its influence on the wake structure.

THE MOVING BELT METHOD

The moving belt replicates the fact that there is no relative motion between the road and the air. The floor boundary layer has to be fully removed by suction just upstream of the belt. In practice, there are several problems with the use of a moving belt. Firstly, care has to be taken to ensure that the belt runs absolutely flat and with little vibration. The belt is normally supported on a flat plate and kept in contact with it by means of suction through air channels or holes in the plate. The suction has to be carefully adjusted; too much suction causes excessive friction and heating, and too little causes the belt to billow upwards or to flap. Breakage of the belt can destroy an expensive model. The model cannot be mounted from underneath, and must be supported on a sting, either from behind or from above. In the latter case there is inevitably some interference between the supporting arm and the flow. If the wheels are in rolling contact with the floor, they will produce a rolling resistance, which must be separated from the aerodynamic drag. It is also difficult to prevent the vertical reaction between the wheels and the floor from upsetting the lift measurements. To overcome these problems, the wheels are normally mounted independently from the body, creating further interference effects.

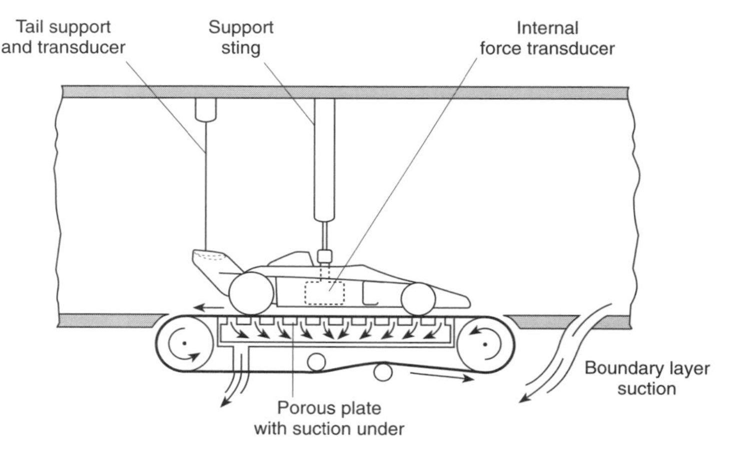

Fig. 11.5 The moving belt method
The suction box and porous plate keep the belt flat. The model has to be suspended from above.

For many years, rival researchers produced data to support claims for or against the use of ground boards in comparison with moving belts. The cynical observer might have been forgiven for noting that those who argued most strongly for the moving belt invariably came from institutions that had invested heavily in the latter. Nowadays, it is generally considered that ground boards give results of moderate accuracy for vehicles such as family cars or commercial vehicles, where the ground clearance is relatively large, but are unsatisfactory for racing and sports cars with small ground clearance. Lajos *et al.*,[23] however, show that even on a high-clearance vehicle such as a bus, ground board tests may not give an accurate representation of all the flow features, particularly in the wake and underside regions.

Figure 11.6 shows a typical experimental set-up for a racing car investigation. The supporting sting and independent wheel mountings may be seen. It is possible to measure the drag on the wheels when they are in contact with the belt, but not the lift. If the wheels are raised just above the floor, then they have to be motorized, which adds complication, and as described in Chapter 2, even a small amount of clearance can strongly affect the wheel lift.

Testing using a moving belt is a complicated and expensive method, with considerable scope for introducing errors. Theoretically, if the tests are conducted with sufficient care and skill, the results should give a good representation of road driving. In the case of ground effect racing cars, the expense and trouble are fully justified, as it is impossible to obtain satisfactory results with a simple ground board. In general, it is found that for racing cars, the fixed ground board method gives reasonably reliable values for the drag coefficient, but is unreliable in terms of lift forces and lift distribution. The problem with lift, particularly on low-clearance racing cars, is that the underside flow and pressure distribution affect it strongly, as

Fig. 11.6 A racing car model arrangement on a moving belt
Note the separate mounting and strain-gauged force-measurement arms for the wheels. The main force-measuring transducers are located inside the model, and attached to the rigid supporting arm. The rear end is suspended by a wire attached to an external force transducer.

described in Chapter 6. The moving belt method will clearly give a much better representation of the under-surface flow.

Experimental studies of the effects of using a moving belt are described in various papers.[2, 3, 7, 12, 15, 19, 20, 22, 23, 26, 27, 34, 37, 38] The oldest of these references dates from 1985, but some nine years later in 1994, there were six papers on the effectiveness of moving belt simulations in the Royal Aero. Society Conference at Loughborough. In one of these papers, Howell[20], despite finding that the moving ground board can produce significantly different results compared with a fixed ground, nevertheless argues that

> *where comparative testing, as in the early development of passenger cars where the criteria are: to generate shapes for minimum drag, to avoid critical geometries, and to understand trends available from fine tuning of the shape, a fixed ground is adequate for assessing underbody shapes.*

THE POROUS FLOOR AND SUCTION METHOD

Another method of floor boundary layer control is to use a porous floor suction, as illustrated in Fig. 11.7. The floor boundary layer is removed by suction through a slot just upstream of the working section. A new layer is prevented from growing by means of the suction from the porous floor section. Experimental investigations

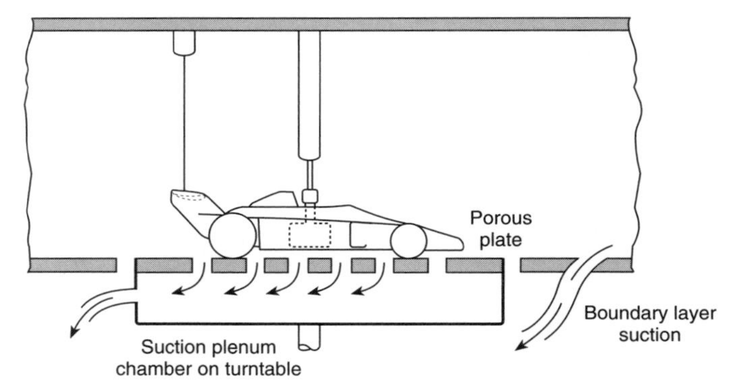

Fig. 11.7 The porous floor and suction method
The lift measurements are rather sensitive to the amount of suction used.

are described by Wulf.[40] The porous floor technique has been adopted by Porsche, and from the work of Eckert *et al.*,[16] it appears that it is possible to obtain aerodynamic load and moment values close to those obtained with the moving belt method, and also, more importantly, close to measurements made on full-scale road tests. The forces are however sensitive to the amount of suction applied, and the difficulty with the technique lies in how to determine the exact amount of suction required. Not surprisingly, since the degree of suction will directly affect the pressure under the vehicle, the lift force is more sensitive to errors in suction than the drag.

Advantages of the method are that it is mechanically simpler and more robust than the moving belt; also, the model can easily be yawed to the wind direction, something that is difficult with a moving belt, because the whole belt assembly has to be swung round. Carr[9] and Carr and Eckert[10] also provide a comparison between moving belt and suction techniques.

THE MIRROR IMAGE METHOD

A further method is to suspend the model in the tunnel with a mirror image underneath as shown in Fig. 11.8. Because of symmetry, the axis of the line dividing the two models must be a streamline, at least in time-averaged flow. There is no floor boundary layer effect. One obvious disadvantage of this method is the cost of making two exactly similar models. A more important objection however is that the flow is not necessarily symmetrical at any instant. Around a road vehicle, there are many areas of turbulent separated flow, and significant random or periodic oscillations occur; thus at any instant, the axis of symmetry is not necessarily a streamline. The method works well for highly streamlined objects such as aerofoils, but is rarely if ever used for automotive vehicle testing nowadays.

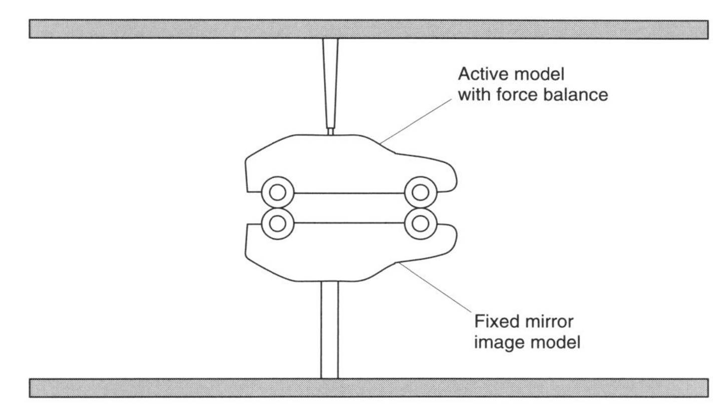

Fig. 11.8 The mirror image method, which is now seldom if ever used.

PRACTICAL COMPROMISES

Despite the apparent importance of using one of the above methods of removing the floor boundary layer, in full-scale tests on anything other than a competition car, it is common practice to simply mount the vehicle on the floor of the tunnel. This is quite justified when comparative rather than absolute measurements are required, and avoids the practical and cost problems associated with lifting a large heavy vehicle, and mounting it on an adequately robust support system. Sucking the boundary layer away just upstream of the vehicle should have much the same effect as mounting the vehicle on a ground board, but even this precaution is not always taken. Academic researchers tend to get carried away with a search for perfection that is not required in the real world of motor manufacturing.

BLOCKAGE EFFECTS

When a model is placed in a closed working section wind-tunnel, the constraining effect of the walls distorts the streamline pattern. In addition, the model partially blocks the tunnel, and the average air speed in the vicinity is therefore increased. The wake behind the vehicle increases the blockage effect. Because the air is speeded up around the model, the forces become larger, and if no correction is applied, the drag and lift coefficients will be overestimated. In addition, the variation of speed with distance along the tunnel is associated with a corresponding streamwise variation in static pressure within the tunnel, and just as the increase in pressure with depth in a fluid tends to cause objects to float upwards, the lengthwise increase in pressure creates a *longitudinal buoyancy* effect which changes the apparent drag.

BLOCKAGE CORRECTIONS

A very crude blockage correction can be obtained based purely on the fact that the same mass of air per second must flow through all parts of the tunnel. This is expressed by the equation of continuity:

$$\rho \times \text{area} \times \text{speed} = \text{a constant}$$

Since the air density does not change significantly at low speeds, it may be taken that the volume of air flowing through all areas is constant:

$$\text{area} \times \text{speed} = \text{a constant}$$

If the tunnel working section area is S and the projected frontal area of the model is A, then around the model, the air has to flow through the remaining area $S - A$. Thus

$$V(\text{true}) \times (S - A) = V(\text{indicated}) \times S$$

where V(indicated) is the speed in the working section upstream of the influence of the model.
V(true) is the actual speed around the model.

$$C_D = \frac{D}{\frac{1}{2}\rho V(\text{true})^2 \times A}$$

$$= \frac{D}{\frac{1}{2}\rho V(\text{indicated})^2 (S/(S - A))^2 \times A}$$

$$= C_D(\text{indicated})((S - A)/S)^2$$

If the blockage is small, second-order terms can be ignored, and it can be written that

$$C_D = C_D(\text{indicated})(1 - 2A/S) \qquad [11.1]$$

There are a number of false assumptions in the simple continuity-based correction given above; the flow is not in reality speeded up by the same amount at all positions around the model, and the speed increase has been calculated on the basis of the maximum cross-sectional area of the model. Clearly for a very streamlined body like an aerofoil, the flow gradually speeds up towards the thickest part, and slows down again towards the rear as the blocking area decreases. In fact, for a streamlined shape, the blockage correction factor for C_D is nearer

$$(1 - 0.5A/S)$$

For a very bluff shape such as a flat plate held normal to the flow, the above expression underestimates the blockage, because a wake is formed that is wider than the model. From the theory of Maskell,[24] a better approximate correction factor for a very bluff shape would be

$$(1 - 2.5A/S)$$

Unfortunately cars are neither completely streamlined nor are they totally bluff. The correction factor should therefore lie somewhere within the wide range of the latter two expressions. A much more sophisticated blockage correction expression is given by Mercker,[25] who separates the solid and wake blockage contributions. Mercker also found that as a first approximation, a modification of the simple momentum-based expression can be used, namely

$$(1 - A/S)^{1.288}$$

This approximates to $(1 - 1.3A/S)$.

Other blockage correction methods that are commonly used are those due to Wuest,[39] and Bettes and Kelly.[4]

The more complicated blockage correction methods take account of the fact that the wide wake produced by flow separation causes an additional blockage effect, and the tunnel walls influence the shape, strength and positioning of the trailing vortices. In general, wall proximity works like ground proximity and weakens the influence of trailing vortices, thereby affecting both the lift and drag measurements.

Although the more sophisticated methods provide better results than the simple continuity-based expression given above as equation [11.1], they do not all produce exactly the same result, and this can lead to apparent discrepancies between tunnels using different methods. Equation [11.1] has the advantage of simplicity, and is commonly used by industry. As long as the area blockage ratio is not too large, the correction applied is in any case small. A medium-sized car in the large MIRA tunnel will have an area blockage ratio of only 5.4 per cent, giving a correction factor of around 11 per cent using expression [11.1].

No correction factor, however sophisticated, can really take full account of the distortions of the flow that the constraints of the walls impose, and it is always advisable to use a small area blockage ratio: ideally less than about 5 per cent.

OTHER METHODS OF CORRECTION

As an alternative to applying correction factors, the profile shape of the wind-tunnel working section walls and roof can be modified to fit the shape of the streamlines of unconstrained flow. Stafford[35] describes a method of calculating the necessary shape using a computational panel method. The obvious problem with this method is that the tunnel shape would need to be tailored to suit each different car model. In practice most cars are of broadly similar shape, so that if the models are all of about the same size, a fixed shape of tunnel can be used with relatively small errors.

OPEN-JET AND SLOTTED-WALL TUNNELS

Figure 11.9 shows an open-jet type of working section. The constraining effect of the walls is removed. The situation is obviously not quite like truly free flow, as there are effects at the boundary between the jet and the still air surrounding it. The blockage correction required is much smaller than for a closed-section tunnel, and usually of opposite sign; that is, the tunnel tends to give apparent drag figures that are slightly too small. Open-jet tunnels also tend to produce an unsteadiness in the flow and distortions of the flow direction.

An alternative to the open-jet tunnel is to use a slotted-wall tunnel, one version of which is shown in Fig. 11.10. The working section walls and roof are not continuous surfaces, but are made up from a series of slats with gaps or slots between, as in duckboarding, or a dense picket fence. The slotted section is surrounded by a plenum chamber. This type of tunnel overcomes most of the problems of unsteadiness and flow direction, and if the solidity of the walls (ratio of closed to open area) is suitably chosen, the blockage errors are small. Early slotted-wall tunnels had the slots running parallel with the flow. Kramer *et al.*[21] describe comparative tests on open-jet and slotted-wall tunnels of this type, and discuss the relative merits. A description of experiments in the large Chrysler slotted-wall tunnel is given by Goenka.[18]

In a more recent variant described by Parkinson,[31] the slats are in the form of aerofoil sections, and run around the working section, as shown in in Fig. 11.10. In this figure, slats may be seen on the walls and roof of the working section. As with the horizontally slotted wall, a plenum chamber is placed around the slotted section to maintain a constant pressure outside the slotted wall. Results presented in Parkinson[31] indicate that by correct selection of the solidity of the wall, an almost negligible blockage effect is obtained. The shaped slats provide some guidance for the edge streamlines without imposing a total constraint.

Fig. 11.9 An open jet wind-tunnel with a model mounted over a ground board.

Fig. 11.10 A circumferentially slotted wall working section, viewed looking along the interior of the tunnel. The slots and slats can be seen on both the walls and the roof of the working section. This photgraph shows the tunnel set-up for investigating wind effects on buildings. The rectangular elements on the floor are used to help generate a simulated atmospheric boundary layer. A similar technique can be used for looking at altmospheric wind effects on vehicles.
(*Photograph courtesy of Paul Blackmore; © Crown copyright Building Research Establishment*)

COMPARATIVE TESTS

Cooper *et al.*[13] reported on an international programme of comparative tests on standard truck models conducted in a variety of tunnels of different type. The same model was used, yet there are quite significant differences in the results obtained, as may be seen from Fig. 11.11. Differences of up to 15 per cent were found in the mean drag coefficient even after blockage correction. Some of the tunnels used in this set of tests were quite small, however, the MIRA 1/4 scale tunnel having a working section of only 1 m by 2.1 m (3.3 ft by 6.9 ft). Relatively close agreement was found between the results of the two larger tunnels.

There have been a number of other international collaborative programmes, using standard test models that were passed around between the collaborating laboratories (see Cogotti *et al.*,[11] Costelli *et al.*,[14] Buchheim *et al.*[5,6] and Carr[8]). In all of these tests significant differences were revealed, particularly between open-jet and closed working section tunnels. In Buchheim *et al.*[6] the maximum discrepancy

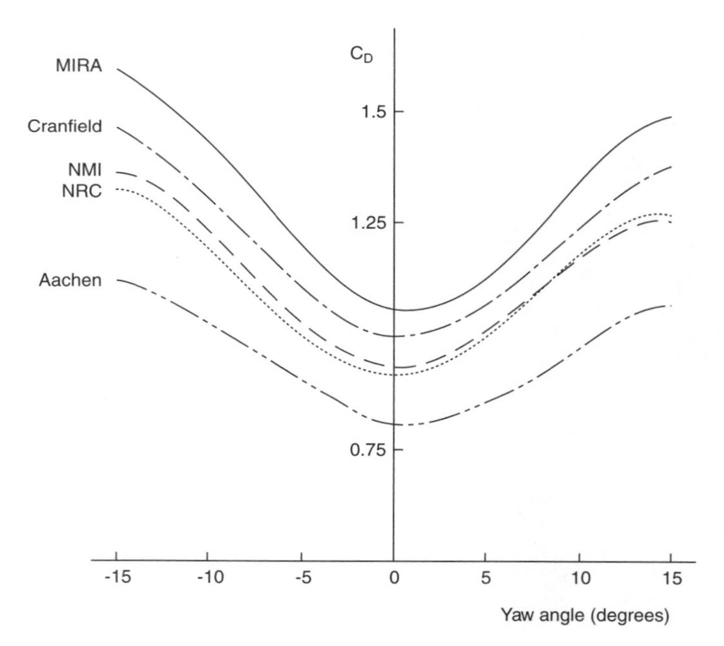

Fig. 11.11 Sample results from comparative wind-tunnel tests on a large truck model. The tunnels used were all relatively small model-testing facilities, not full-scale tunnels. (*After Cooper et al.*[13])

The institutions represented at the time were:

- MIRA — Motor Industry Research Association (UK) model tunnel
- Cranfield — Cranfield College of Aeronautics (UK)
- NMI — National Maritime Institute (UK)
- NRC — National Research Council (Canada)
- Aachen — Fachhochschule Aachen (Germany)

between drag coefficient values was as high as 9.3 per cent for one vehicle configuration. The lift force measurements showed even greater inconsistencies than drag measurements. In some cases discrepancies have been attributed to the measurement systems used, particularly the method of mounting the model and of measuring the forces. One of the most important factors is the choice of position for measuring the tunnel static pressure, particularly in the case of open-jet tunnels. The design of floor-mounted load-measuring pads was also found to be an important factor.

Some caution has to be exercised when looking at comparative tests. Unless these are conducted blindly with no prior knowledge of the expected results, experimenters may sometimes be inclined to select the most convincing data for presentation, or may adjust blockage correction factors to optimize the fit.

In addition to comparison between different wind-tunnels, comparisons have been made between wind-tunnel and road tests, as in Costelli.[14] Comparative studies, and particularly blind tests, enable the laboratories concerned to refine their blockage correction factors and test procedures. Testing practices are gradually becoming standardized internationally, which helps to remove discrepancies. A reputable wind-tunnel laboratory should nowadays be able to produce results that are accurate to within a few percentage points.

IMPORTANCE OF MODELLING FINE DETAIL AND INTERNAL FLOWS

One difficulty with wind-tunnel modelling is the necessity to model fine detail and internal flows accurately. Olson and Schaub[29] have made a careful study of such effects for the case of a large truck model. Accurate and approximate half-scale models were used. Engine cooling air flow was found to be responsible for 3.8 per cent of the total drag at a Reynolds number of 2.2×10^6. The simpler block-form model was found to have a drag coefficient of up to 11 per cent higher than the accurate model in the short version of the vehicle, though with only a 2 per cent error for a longer version. These findings show the importance of conducting extensive tests before drawing general conclusions. Had only the long-form truck been tested, it might have been concluded that accurate modelling only produced a marginal effect. The 11 per cent difference on the short-form vehicle is a large error. Detailed fine modelling makes wind-tunnel models extremely expensive.

ROAD TESTING

The principal difficulty in road measurements lies in separating the contributions due to rolling resistance and mechanical friction from the aerodynamic drag. The mechanical friction and rolling resistance can be measured by placing the vehicle on rollers, but tyre rolling resistance is sensitive to surface material and texture and to the radius of curvature of the rollers. Drums used for standard tyre tests are normally smooth steel, and this gives nothing like the characteristics of contact with a real road.

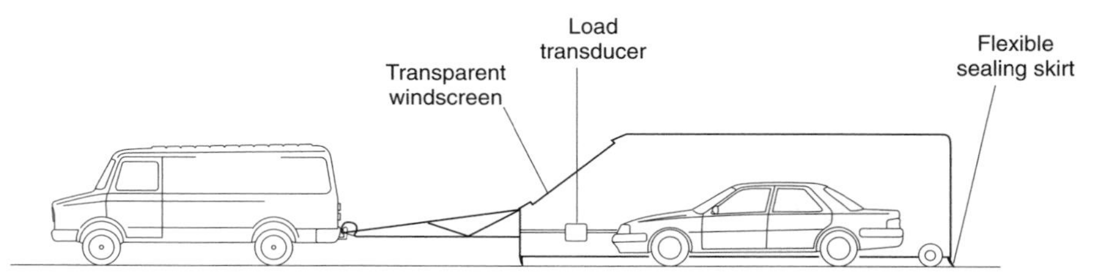

Fig. 11.12 The use of a towing shroud to eliminate aerodynamic drag from rolling resistance measurements.

One method of separating the rolling and mechanical resistance from the aerodynamic drag is to tow the vehicle inside a shroud as illustrated in Fig. 11.12. The shroud fits closely to the road, and is almost sealed with a skirt. This removes any aerodynamic drag effects on the vehicle. The tension force in the tow bar connecting between the shroud and the towed vehicle gives the drag due to tyres and mechanical friction.

The total drag is usually measured by coast-down tests. The vehicle is driven up to a suitable speed; the gearbox is then put into neutral, and the car is allowed to coast down to a predetermined lower speed. From measurements of distance, time and speeds at start and finish, it is possible to infer the total resistance. A number of repeat runs are made in both directions to eliminate the effects of any slope or wind. The previously measured rolling and mechanical friction resistance can then be subtracted to give the aerodynamic drag. In practice, the methods used are much more complicated than in the simple description given above. Morelli *et al.*[28] give details of the procedures and precautions. Discrepancies in drag coefficient figures between wind-tunnel and road tests reported in this reference vary from 0.22 per cent to 1.56 per cent. Roussillon[33] and Costelli *et al.*[14] report similar experiments.

For lift measurements, the vehicle can be driven over a stiff weighbridge plate set flush with the road, as described by Eckert *et al.*[16] The results quoted in the references indicate that when adequate care is taken, there is normally a close correlation between road and wind-tunnel tests.

MEASUREMENT METHODS FOR AUTOMOTIVE WIND-TUNNEL TESTING

An adequate description of wind-tunnel test techniques would and indeed does fill complete books. This chapter will therefore be restricted to a description of the special measurement techniques required for road vehicle testing. For general details of wind-tunnel testing techniques, Pope[32] and Pankhurst and Holder[30] are highly recommended.

MOUNTING AND FORCE MEASUREMENT

Car models can be mounted in the same way as aircraft models: that is, either on a set of supports connected to an external force-measuring system, or on a rigid sting with an internal miniature multiple-axis force-measuring transducer. Overall forces and moments can be measured in this way, and from the pitching moment and lift force measurements, the distribution of aerodynamic load between the front and rear wheels can be determined. Another method normally employed for full-scale cars is to place the wheels on measuring pads set into the floor. Care has to be taken in order to avoid a large pad area being exposed to the local tunnel pressure field, otherwise false readings can be obtained. This method cannot of course be used in conjunction with a moving belt system.

PRESSURE MEASUREMENT

On small models the local surface pressure can be measured by drilling a small hole or pressure tapping through the surface. A tube is connected to the hole at one end, and to a pressure-measuring device at the other: nowadays normally a pressure transducer, which produces an output voltage proportional to the applied pressure. Because of the relatively small changes in pressure that have to be measured, the transducer has to be very sensitive, and until recently such transducers were expensive. A large number of tappings may be needed, and to avoid expense it is common to use a scanning valve which connects each of a number of pressure tappings in turn to a single transducer. The falling cost of transducers and the advantages of being able to make pressure measurements simultaneously have recently encouraged laboratories to invest in large arrays of individual transducers instead of scanning valves.

On full-scale cars, drilling pressure tappings would represent a highly destructive form of testing, and as an alternative, very small static pressure probes only 0.5 mm (0.02 in) thick can be taped directly on to the external surface of the vehicle. The technique will slightly affect the boundary layer flow, but generally to an acceptable degree.

THE ROTATING WHEEL PROBLEM

On a fixed ground board the wheels obviously cannot be rotated whilst still in contact with the road. Rotating wheel models can be used, but the clearance gap required will adversely affect the lift measurements. Stapleford and Carr[36] found that by chance, the values of C_L and C_D given by rotating wheels with zero ground clearance (obtained by extrapolation from very small clearances) were also given approximately by fixed wheels at a small ground clearance corresponding to about 1.5–3 per cent of the wheel diameter, as may be seen in Fig. 11.13.

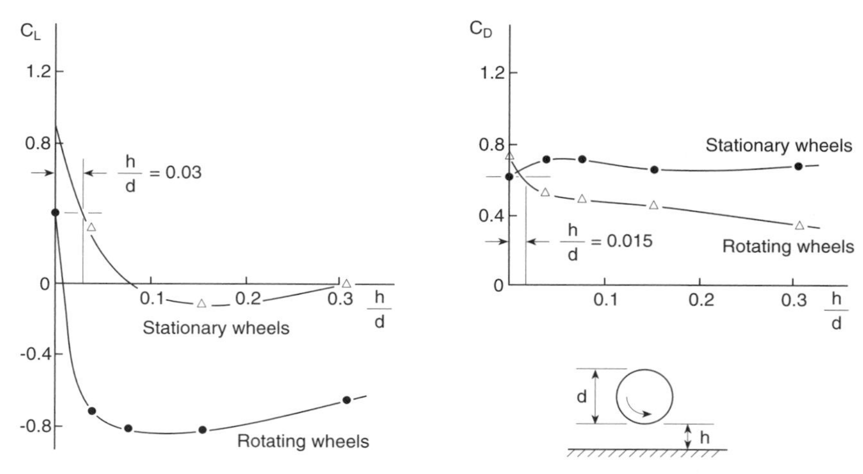

Fig. 11.13 The effect of ground proximity on the lift and drag of rotating and fixed wheels. At a small clearance, the forces on the stationary wheel are the same as those (estimated) for a rotating wheel in contact with the road.
(*After Stapleford and Carr*[36])

When a moving belt floor is used, the wheels can be rotated at the correct speed by contact with the floor. The drag due to rolling and mechanical friction is assessed by running the belt with the wind turned off. Unfortunately this may change during the run, and the wind influence on the belt can affect the results. The largest problem, however, particularly in the case of exposed-wheel racing cars, is that the total lift force cannot be measured directly when the wheels are in contact with the road, because the lift and weight forces are balanced by the reaction force from the road. The lift force on the bodywork can be assessed in the normal way if the wheels are detached and separately mounted, as shown in Fig. 11.7. The wheels in this case have separate transducers for measuring the drag on the wheels, but wheel lift still cannot be measured. An alternative method of separating the wheel and body lift forces is to attach the wheels to the car by a low-friction suspension system, but the method is not foolproof, and vertical force components from the wheels can be transmitted to the body.

FLOW VISUALIZATION

Articles on vehicle aerodynamics in the popular press are invariably accompanied by photographs of smoke streams around a vehicle. Smoke visualization does have its uses, but they are limited. Smoke is useful in identifying areas of separation, reattachment and recirculation. Figure 11.14 shows a hand-held smoke-emitting wand being used to show the centre-line flow over an open cabriolet in the full-scale wind-tunnel at MIRA. Finer details can sometimes be discerned by using a powerful laser to illuminate smoke in a thin sheet of flow. The problem with smoke is that it mixes very rapidly in the turbulent flow regions that are of particular interest, and in the case of a closed-return type of tunnel, the whole tunnel rapidly fills with smoke. Open-return tunnels do not suffer from this problem, but exhaust into the surrounding room making the local environment unpleasant.

Painting the vehicle surface with a fine powder suspended in oil or paraffin (kerosene) also provides information on the location of transition, separation and reattachment lines. When the wind is applied, the suspension is scoured away rapidly in regions of attached flow but accumulates in regions of separated flow, the separation line showing up as a distinct change in pigment density. Some experience is necessary when interpreting the results of this technique.

Small tufts of wool attached to the vehicle surface give an indication of regions of separation, attachment, reversed flow, vortical flows and high turbulence, together with an indication of the local flow direction. A small wool tuft on the end of a long wand provides a useful means of manually traversing the flow to identify similar flow features.

Fig. 11.14 A hand-held smoke-emitting wand being used to show the centre-line flow over an open cabriolet in the full-scale wind tunnel at MIRA. Note the well-attached flow up to the top of the front screen.
(*Photograph courtesy of J. Howell, Rover Group Ltd*)

FEATURES OF THE AUTOMOTIVE WIND-TUNNEL

From the foregoing, it may be seen that a dedicated automotive wind-tunnel has a number of features that distinguish it from an aeronautical tunnel:

- It should ideally be large enough to accommodate a full-size road vehicle.
- A moving belt should be provided.
- A circumferentially slotted working section wall appears to offer the best solution to the problem of blockage.

Apart from measuring lift and drag forces, wind-tunnels are also used to assess the performance of vehicle systems in adverse environment conditions such as extremely low or high temperatures. Such tunnels need not be as large as the type used for force measurements, and it is common practice to use a separate environmental wind tunnel.

Sound measurements in and around cars are also of importance, and for this purpose a quiet tunnel with acoustic lining to inhibit reflected sound waves is required.

REFERENCES

1. Barnard R. H., The aerodynamic tuning of a group C sports car, *Journal of Wind Engineering and Industrial Aerodynamics*, Vol. 22, Elsevier Science Publications, 1986, pp. 279–89.
2. Bearman P. W., DeBeer D., Hamidy E. and Harvey J. K., The effect of moving floor on wind-tunnel simulation of road vehicles, SAE paper 880245, Detroit, 1988.
3. Berndtsson A., Eckert W. and Mercker E., The effect of ground plane boundary layer control on automotive testing in a wind tunnel, SAE paper 880248, Detroit, 1988.
4. Bettes, W. H. and Kelley K. B., The influence of wind tunnel solid boundaries on automotive test data, SAE paper 741031, 1975.
5. Buchheim R. *et al.*, Comparison tests between major European automotive wind tunnels, SAE Congress paper 800140, Detroit, February 1980.
6. Buchheim R. *et al.*, Comparison tests between major European and North American automotive wind tunnels, SAE paper 830301, Detroit, 1983.
7. Burgin K., Adey P. C. and Beatham J. P., Wind tunnel tests on road vehicle models using a moving belt simulation of ground effect, *Journal of Wind Engineering and Industrial Aerodynamics*, Vol. 22, pp. 227–36, 1986.
8. Carr G. W., Correlation of aerodynamic force measurements in MIRA and other automotive wind tunnels, SAE paper 820374, 1982.
9. Carr G. W., A comparison of the ground-plane-suction and moving-belt ground-representation techniques, SAE paper 880249, Detroit, 1988.
10. Carr G. W. and Eckert W., A further evaluation of the ground plane suction method for ground simulation in automotive wind tunnels, SAE paper 940418, 1994.
11. Cogotti A. *et al.*, Comparison test between some full-scale European automotive wind tunnels – Pininfarina reference car, SAE congress paper 800139, Detroit, 1980.
12. Cooper K. R. and Fediw A. A., Development of a moving ground belt system for the study of vehicle aerodynamics, *Proc. Road Vehicle Aerodynamics Conference*, RAeS, Loughborough, July 1994, pp. 9.1–11.

13. Cooper K. R., Gerhardt H. J., Whitbread R., Garry K. P. and Carr G. W., Comparison of aerodynamic drag measurements on model trucks in closed-jet and open-jet wind tunnels, *Proc. 6th Colloquium on Industrial Aerodynamics*, Aachen, June 1985, pp. 261–74.
14. Costelli A. *et al.*, Fiat Research Centre reference car: Correlation test between four full scale European wind tunnels and road, SAE paper 810187, Detroit, 1981.
15. Diuzet M., The moving belt of the I.A.T. long test-section wind tunnel, *Proc. 6th Colloquium on Industrial Aerodynamics*, Aachen, June 1985, pp. 159–68.
16. Eckert W., Singer N. and Vagt J. D., The Porsche wind-tunnel floor-boundary layer control – a comparison with road data and results from moving belt, *Vehicle Aerodynamics, Wake Flows, Computational Fluid Dynamics, and Aerodynamic Testing*, SAE SP-908, 1992.
17. Garry K. P., Wind tunnel tests on the influence of fixed ground board length on the aerodynamic characteristics of simple commercial vehicle models, *Journal of Wind Engineering and Industrial Aerodynamics*, Elsevier Science, Vol. 38, 1991, pp. 1–10.
18. Goenka L. N., Scale-model tests on the test section of the Chrysler slotted wall automotive wind-tunnel, Vehicle Aerodynamics, Recent Progress, SAE SP-855 paper No. 910313, 1992.
19. Hackett J. E., Baker J. B., Williams J. E. and Wallis S. B., On the influence of ground movement and wheel rotation on modern car shapes, SAE paper 870245, Detroit, 1987.
20. Howell J., The influence of ground simulation on the aerodynamics of simple car shapes with an underfloor diffuser, *Proc. Road Vehicle Aerodynamics Conference*, RAeS, Loughborough, July 1994, pp. 36.1–11.
21. Kramer C., Gerhardt H. J. and Janssen L. J., Flow studies of an open jet wind tunnel and comparison with closed and slotted walls, *Proc. 6th Colloquium on Industrial Aerodynamics*, Aachen, June 1985, pp. 7–28.
22. Lajos T. and Hegel I., Some experiences of ground simulation with moving belt, *Proc. Road Vehicle Aerodynamics Conference*, RAeS, Loughborough, July 1994, pp. 34.1–11.
23. Lajos T., Preszler L. and Finta L., Effect of moving ground simulation on the flow past bus models, *Proc. 6th Colloquium on Industrial Aerodynamics*, Aachen, June 1985, pp. 227–36.
24. Maskell E. C., A theory of blockage effects on bluff bodies and stalled wings in a closed wind tunnel, Ministry of Aviation (UK), Reports and Memoranda R & M 340, November 1963.
25. Mercker E., A blockage correction for automotive testing in a closed wind tunnel, *Proc. 6th Colloquium on Industrial Aerodynamics*, Aachen, June 1985, pp. 41–54.
26. Mercker E. and Knape H. W., Ground simulation with moving belt and tangential blowing for full-scale automotive testing in a wind tunnel, SAE paper 890367, Detroit, 1989.
27. Mercker E., Soja H. and Wiedermann J., Experimental investigation of the influence of various ground simulation techniques on a passenger car, *Proc. Road Vehicle Aerodynamics Conference*, RAeS, Loughborough, July 1994, pp. 10.1–11.
28. Morelli A., Nuccio P. and Visconti A., Automobile aerodynamic drag on the road compared with wind tunnel tests, SAE paper 80186, 1981.
29. Olson M. E. and Schaub U. W., Aerodynamics of trucks in wind tunnels: the importance of replicating model form, model detail, cooling system and test conditions. Vehicle Aerodynamics, Wake Flows, Computational Fluid Dynamics, and Aerodynamic Testing, SAE SP-908, paper 920345, 1992.
30. Pankhurst R. C. and Holder D. W., *Wind Tunnel Technique*, Pitman, London, 1965.
31. Parkinson G. V., Advantages of a passive low-correction wind tunnel, *Proc. Inaugural Conference of the Wind Engineering Society*, Cambridge, September 1992.
32. Pope A., *Wind Tunnel Testing*, 2nd edn, Wiley, New York, 1964.
33. Roussillon G., Contribution to an accurate measurement of aerodynamic drag from coast-down test and determination of actual rolling resistance, *Proc. 4th Colloquium on Industrial Aerodynamics*, Aachen, June 1980, pp. 53–74.

34. Sardou M., The sensitivity of wind-tunnel data to a high speed moving ground for different types of road vehicles, SAE 880246, 1988.
35. Stafford L. G., A streamline wind tunnel for testing at high blockage ratio, *Proc. 4th Colloquium on Industrial Aerodynamics*, Aachen, June 1980, pp. 35–52.
36. Stapleford W. R. and Carr G. W., Aerodynamic characteristics of exposed rotating wheels, MIRA report No 1970/2.
37. Wiedermann J., Some basic investigations into the principles of ground simulation techniques in automotive wind tunnels, SAE paper 890369, Detroit, 1989.
38. Wildi J., Wind tunnel testing of racing cars – the importance of the road simulation technique, *Proc. Road Vehicle Aerodynamics Conference*, RAeS, Loughborough, July 1994, pp. 11.1–12.
39. Wuest W., *Strömungsmesstechnik*, Vieweg, Braunschweig, 1969.
40. Wulf R., Investigations on a plate with uniform boundary layer suction for ground effects in the 3 m by 3 m low speed wind tunnel of DFVLR-AVA, AGARD-CP-174, October 1975.

CHAPTER 12

COMPUTATIONAL FLUID DYNAMICS (CFD) METHODS

THE NUMERICAL APPROACH

It has long been a dream of engineers that they might one day be able to calculate the aerodynamic forces on vehicles in much the same way that they can calculate structural stresses and deflections. By the outbreak of the Second World War, considerable advances had been made in the analysis of flows around aircraft, but cars presented an apparently insurmountable problem. This is because there are two major differences between the type of flows involved. Firstly, aircraft flows are fully attached over most of the surface, and secondly, they tend to be either axisymmetric (as on the cylindrical fuselage of an airliner) or nearly two-dimensional (as on much of the wing and tail surfaces). On cars, the flows are strongly three-dimensional, and the flow field is dominated by the effects of separation. These factors prevent the use of many of the simplifying assumptions applied to aircraft, and it was not until the advent of the digital computer that calculating the flow field around road vehicles became a realistic possibility.

To determine the aerodynamic forces on a vehicle, it is first necessary to calculate the pressures and flow velocities around the surface, which can be done by making use of the equations of motion of a fluid. These are usually expressed in the form of a set of partial differential equations. The early methods of analysis, developed largely for aeronautical applications, involved choosing simple body contour shapes and making simplifying assumptions which eventually produced equations connecting the pressure or velocity at any point in space to its coordinates. Only certain shapes could be handled, but these did include aerofoil forms. The advent of the computer, however, opened up a new possibility: instead of having to solve the differential equations by mathematical analysis, they could be solved directly by numerical means.

As an example of the difference between an analytical and a numerical solution, consider the problem of finding the area under a line following a simple relationship such as $y = x^2$, as illustrated in Fig. 12.1. The area can be determined by an analytical method: that is, x^2 can be integrated over the appropriate limits, which in the case illustrated, are 0 and 1, giving 1/3 square units as the answer.

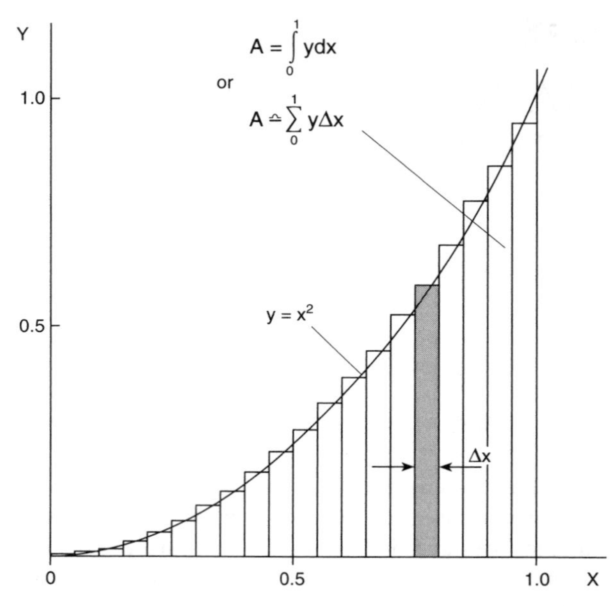

Fig. 12.1 Analytical and numerical solutions to the problem of determining the area under a line between the limits $x = 0$ and $x = 1$.

The alternative *numerical* approach is to split the area under the curve into narrow rectangles, and to work out the area of each by calculating the height and multiplying by the width. The total area under the curve between the specified limits is found by adding the areas of all the rectangles. The latter method will not give exactly 1/3 as the answer, but by making the size of each rectangle smaller and smaller, the accuracy can be increased to any level required. In the case of a much more complicated curve such as the outline of a car, it may not be possible to find a mathematical expression that precisely represents the line, and in this case the numerical approach is required.

Numerical methods were used for solving simple flow problems even before the invention of computers, but as this involved lengthy hand calculations, their use was restricted to a limited number of special applications such as aerofoil section design.

THE EQUATIONS OF FLUID MOTION

The most general set of equations of fluid motion normally used are the so-called **Navier–Stokes equations**. These give the relationship between pressure, momentum and viscous forces in three-dimensional space. Additional equations are the **continuity equation**, which describes the balance between the quantity of fluid flowing into and out of a space (the two must be equal in a steady flow), and the **energy equation**, which relates to the interchange between different forms of energy: potential, kinetic etc., through the flow. When the three sets of equations are combined, they provide the possibility of determining the pressures and velocities anywhere in the flow around any shape. There are some special circumstances where they are not adequate, but that need not concern us here. For road vehicles, the effects of compressibility (air density changes) are negligible, the energy equation is not required, and a simplified 'incompressible' version of the Navier–Stokes equations can be used.

One way of using these equations is to split the flow field into a three-dimensional **grid** of blocks, as shown in Fig. 12.2. At the corners (or the centre) of each block, the values of the various variables, velocity etc., have to be determined. Only when the variables at all points are correct will the equations be satisfied. The variables can be so-called 'primitive' items such as pressure, or if a different formulation of the equations is used, they can be combined quantities such as momentum: the product of mass and velocity. The numerical methods involve systematically predicting and correcting the values at each point iteratively. The true correct values are never found, and the computation is simply stopped when the calculated values cease to alter significantly with each new iteration.

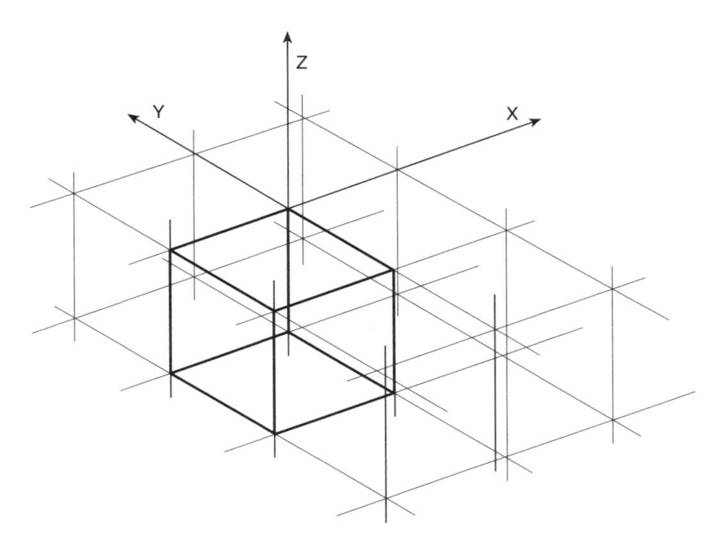

Fig. 12.2 A simple three-dimensional grid system using cube-shaped blocks.

PRACTICAL PROBLEMS INVOLVED IN USING THE NAVIER-STOKES EQUATIONS

The problem with using the full Navier–Stokes equations is that the solution will only match the real physical flow accurately if account is taken of the very finest details of turbulent motion. If you look at turbulent motion, as in the smoke from a fire, you will see that it contains both large eddying motions and small-scale movements. Unfortunately, the smallest scales are microscopic, and to take account of their effect fully, the computational grid has to be even smaller. It has been estimated that to capture the effect of the smallest turbulent motions and solve the flow around a structure the size of a car, a grid of around 10^{18} points would be needed. By the mid 1990s, the largest number of grid points that could be handled was only around 10^{6}, so **direct numerical solution** (DNS) of the Navier–Stokes equations is not a practical proposition for vehicle aerodynamics in the immediate future. To date, the technique has only been used to study very simple cases such as the flow over a flat plate.

SIMPLIFICATIONS: TURBULENCE AND BOUNDARY LAYER MODELLING

To get around the computational problem, it is necessary to simplify the equations by making some reasonable assumptions or approximations. One way is to simplify the terms that describe the turbulent motion. Instead of providing for a full description of the turbulence, a **turbulence model** is substituted. The model only takes account of important average effects of turbulence, such as the rate at which the energy is transported through the fluid, and the rate at which it is dissipated.

As a halfway-house between direct numerical solution and turbulence modelling there are the so-called **large eddy simulation** techniques, where the smaller scales of turbulence are modelled approximately, but larger scales are modelled more directly. Again, prodigious amounts of computing power and time are required even for simple problems, and the technique is unlikely to be of much use in vehicle aerodynamics in the medium term.

In addition to modelling the turbulence properties, the behaviour of the flow in the boundary layer can be modelled approximately on the basis of experimental measurements. For example, it is known that on a smooth surface, in a two-dimensional boundary layer with a weak streamwise variation of pressure, the variation of average flow velocity with distance from the surface is quite well approximated by a simple logarithmic expression. Simple relationships such as this are however not applicable in cases where the flow is highly three-dimensional: that is, where there are strong cross-flows. Some progress in modelling boundary layers in three-dimensional flows has been made, but significant improvements are required.

FURTHER SIMPLIFICATIONS

A much greater simplification of the Navier–Stokes equations is obtained if the viscosity terms are removed. The resulting equations, normally known as Euler's equations, are useful in aeronautical transonic and supersonic flow calculations, but the removal of the viscosity terms means that they cannot reproduce the effects of turbulence or flow separations directly. The effects of turbulence in the boundary layer can be taken account of by a two-stage method, where the flow field outside the boundary layer is calculated using the Euler equations, and then a different set of simplified equations is used to solve the flow in the boundary layer. This approach works well for the large areas of nearly two-dimensional attached flow found around aircraft, but cannot readily be used for the highly three-dimensional and separated flows around road vehicles, and the approach has largely been abandoned by vehicle aerodynamics researchers.

An even simpler set of equations is obtained if in addition to ignoring the viscosity terms, the flow is assumed to be irrotational: that is, it contains no regions where fluid particles are rotating. This might seem a severe limitation, because there is rotation in the boundary layer, and in any separated wake or in the centre of a vortex, but as will be described, there are some ways of getting around these problems, at least partially.

This final simplification of the equations of motion produces what is known as **potential flow**. The potential flow equations can be solved analytically for some special cases, but again computers have greatly extended the class of flow problems that can be solved adequately for engineering purposes. Potential flow solution techniques were developed to a high level of refinement for aircraft applications, where rotational flow is usually confined to the boundary layer, and to a thin sheet behind the trailing edge of the wing.

Methods using the potential flow equation require much less computational capacity and speed than those using either the Euler or the Navier–Stokes equations. In addition, with potential flow methods, it is only necessary to satisfy the equations at a grid of points on the surface of the body, so the grid required is effectively only a two-dimensional net draped over the surface, as illustrated in Fig. 12.3. This simple net reduces the computational demand considerably in comparison with methods requiring a three-dimensional grid.

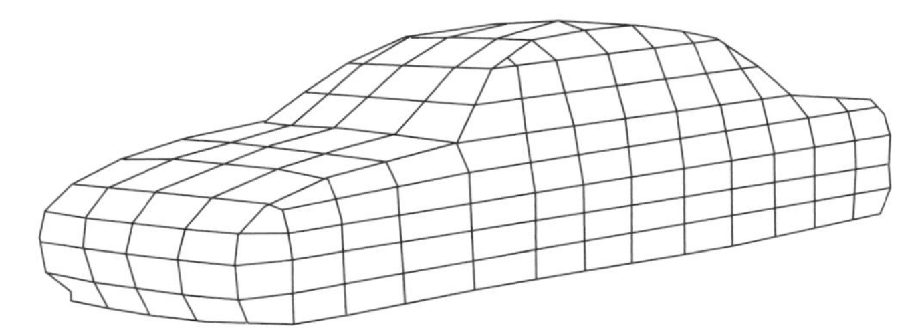

Fig. 12.3 In potential flow panel methods a two-dimensional net is draped over the surface.

PANEL METHODS

Most potential flow solution methods use a large set of imaginary flow elements which are employed at either the edges or the centres of the net cells. Because the elements are applied to panels of the surface or boundary grid, the techniques are called **panel** or **boundary element** methods. The elements may be vortex rings (or hoops), doublets, or sources and sinks. A source is a point through which fluid is emerging, and a sink is one into which it is disappearing; a doublet is a combination of the two. The methods are described in most textbooks on classical fluid mechanics. There are also other potential flow solution methods involving mathematical transformations, but these are now seldom used.

Figure 12.4 shows the surface grid for a panel method using vortex rings. The rings are roughly rectangular. Each part of every vortex ring induces a small increment of velocity at any point in the flow. The resultant velocity at that point is the vector resultant of the free-stream velocity, and all the small induced contributions from all parts of all vortex rings on the whole surface. On the surface of the body, the flow must be parallel to the surface direction, otherwise the air would have to be flowing through it. The strengths of the vortices therefore have to be such that the resultant flow everywhere on the surface is parallel to the surface direction. The numerical solution involves solving a large set of simultaneous equations to determine the vortex strengths, something that computers can do quite

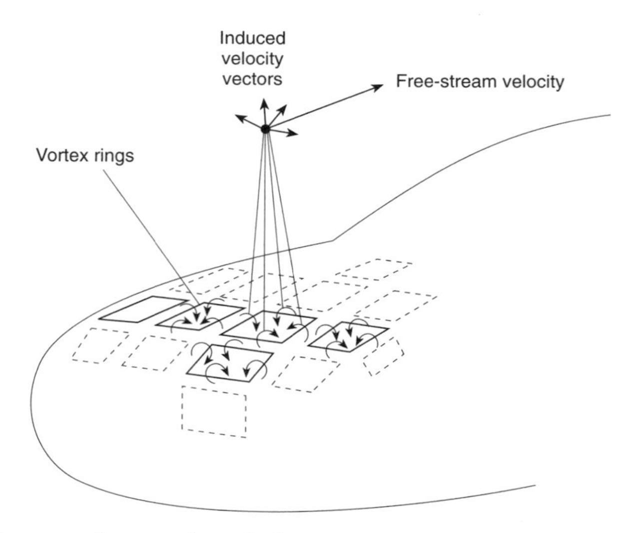

Fig. 12.4 A vortex ring panel method
The surface is covered by a grid of imaginary vortex rings. The velocity at any point in space is the resultant of the free-stream velocity and the velocity components induced by every part of all vortex rings.

quickly. Once the equations have been solved it is easy to calculate the velocity and pressure at any point in the flow field, not just on the surface. In practice it is necessary to add at least one doublet to complete the solution, as described in Stafford.[7]

Following success with panel methods applied to aircraft, attempts were made to adapt the technique to cars. Stafford[7] describes the use of a vortex element method to solve for the flow around a car-shaped body. The results showed apparently good agreement with experimental measurements of the pressure distribution around the front of the vehicle. In these largely attached flow regions, the potential flow equations would be expected to work well, because turbulence and viscosity have little influence. Panel methods cannot directly predict the areas of stable separated flow such as occur around the base of the windscreen, but these do not strongly affect the flow field. At the rear of the vehicle, however, the flow is dominated by a wake in which the potential flow equations cannot be applied because the flow is rotational and the effects of viscosity are important. One way to get around this is to guess the shape of the wake, and replace it by an imaginary solid body of the same shape. Some of the effects of the boundary layer can also be approximated by making the body thicken slightly towards the rear.

Early computational experimenters sometimes went one stage further than guessing the wake shape; they measured it using flow visualization and wind-tunnel models. Not surprisingly, using the measured wake shape, the computational model gave a fairly good agreement with experimental measurements of the pressure distributions. The problem with this approach is that if the computer solution has to rely on wind-tunnel data, then for practical engineering purposes, one might just as well stick with the wind-tunnel and not bother with the computational model. The real disadvantage of the panel method, however, it that since the wake flow is not modelled, merely replaced by a solid boundary, it is impossible to estimate the pressure on the rear of the vehicle accurately; this in turn means that the most important quantity, the drag, cannot be predicted reliably.

As a general method for car applications, the inability to predict and model separated rotational flows means that the potential flow methods are of limited value. There are, however, some situations where they can be useful, including modelling of blockage effects in wind-tunnels, where moderate errors in the computed flow are not too significant, and the design of aerofoil sections, where the flow should normally be attached. They can also be used to find suitable areas for locating air intakes and exits. Potential flow methods can be useful in fundamental investigations; a good example is provided by Katz,[3] who looked at several interference effects on racing car wings such as ground and body proximity, and the effect of using multiple wings (see Chapter 6).

Considerable research and development work was applied to the potential flow methods, including the marriage of potential flow solutions with boundary layer modelling; however, with the advent of well-developed computational packages capable of dealing with a fuller version of the Navier–Stokes equations, the emphasis has now shifted towards methods using the latter.

USING PROGRAMS TO SOLVE THE NAVIER-STOKES EQUATION

Although aircraft aerodynamicists used the inviscid (no viscosity terms) Euler equations to good effect, road vehicle aerodynamicists realized that because of the importance of wake and separated flows, they needed to include the effects of viscosity, so the more complex Navier–Stokes equations would be required. Because of the complexity of the flows, and limitations on the processing power available, only the reduced form of the Navier–Stokes equations with a turbulence model is currently practical.

There are many different techniques of numerical modelling using the Navier–Stokes equations, and almost as many program packages. Some organizations have developed their own computer codes, but commercially available packages are more commonly used. All packages have their advocates, but each has its own pros and cons, and there is no obvious best buy for all applications. The commercial packages have the advantage of including software to help define the geometry of the vehicle and the grid; they can also 'post-process' the computed data to provide graphical output. The attraction of these Navier–Stokes packages is that they are easier to use than some of the specialized aeronautical software, and they can be employed for a wide range of different types of problem. In practice, users have found that a great deal of experience and specialist knowledge is required when using them for automotive applications.

LIMITATIONS OF CURRENT NAVIER-STOKES SOLVERS

Despite the advances in the coding, and experience in running program packages using the Navier–Stokes equations, external flows over cars still represent a major challenge. The primary problems are as follows.

1. Existing popular turbulence models do not work well in all areas of the flow.
2. Modelling the effects of turbulence in the oncoming wind is difficult or expensive in terms of processor running time.
3. Fine details on cars such as hub caps and body-panel joints can have a significant effect on the flow, but to model them sufficiently accurately would require a very fine and hence large grid.
4. Generating the required grid shapes requires many hours of work by skilled users.
5. Good simulations require long run times on very powerful computers.
6. It is difficult to correctly model the rotation of the wheels.
7. The wake behind vehicles is unsteady, and this feeds back to unsteadiness on the flow around the vehicle; time-dependent flow solutions are very expensive in terms of computer run time.
8. Methods of modelling boundary layer behaviour in three-dimensional flows are not well developed, and direct numerical solution is not practical.
9. As a consequence of the above, separation positions are not accurately predicted, and this can considerably reduce the accuracy of the solution.

Despite these problems, researchers have made considerable progress in some areas, as outlined below.

GRIDS

In order to minimize the number of cells used, a great deal of ingenuity is employed. In areas where the flow properties are changing rapidly with respect to position, as in the boundary layer, a very fine mesh of grid points is used, whereas in less sensitive areas, away from the vehicle, a coarser grid is employed (see Fig. 12.5). Some considerable skill is required in order to generate a suitable shape of grid. The commercial packages normally incorporate various aids, and often a so-called automatic grid generation program is included. With time, these routines have become both more sophisticated and easier to use.

A significant problem is that the solution accuracy is influenced not only by the spacing and number of cells in the grid, but also by the shape of the cells. Discontinuities in size and shape all have an effect.

The rapid increase in the memory size and processing speed of computers has enabled finer and finer meshes to be generated. Figure 12.6 for example shows the level of shape detail that can be been modelled using 98 000 cells. Even with double the number of cells in each direction (total 782 000 cells), it is still far from fine enough to correctly model the effects of surface protrusions, and it will take a few more generations of processor before that becomes possible.

One problem with grid generation on road vehicles is that although the overall form may look simple, the small but important details of the geometry are extremely complex. The burden of the computational fluid dynamics (CFD) user may be lightened by transferring the data describing the coordinates of the body from the computer-aided design (CAD) system to the CFD package.

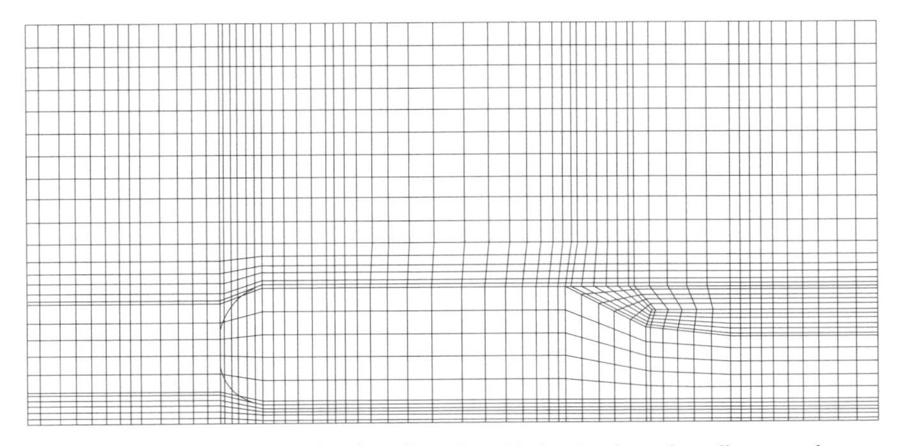

Fig. 12.5 Part of a two-dimensional section of a grid showing how the cells are made smaller in regions of rapid change in flow properties.
(*Courtesy of Dr J. Graysmith, MIRA*)

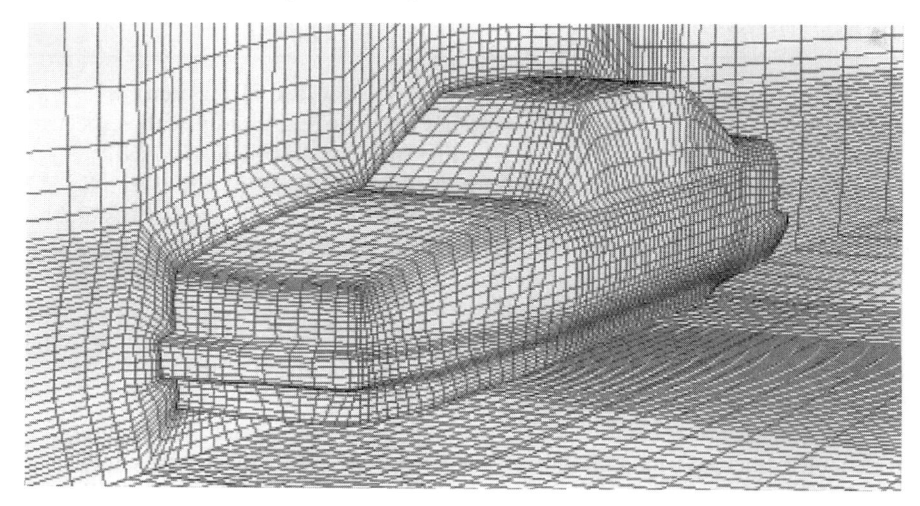

Fig. 12.6 Surface intersection lines of a grid of 98 000 cells on a simplified Volvo 850 car shape. A finer mesh containing double the number of lines in each coordinate direction was used giving a total of 782 000 cells.
(*After Ramnefors et al.,*[5] *courtesy of M. Ramnefors and Volvo Data AB*)

STANDARD TEST CASES

In the early days, some exaggerated claims were made about the accuracy and reliability of computed solutions for cars. In order to provide a standard measure, various 'benchmark' test shapes have been adopted. One of these is the Ahmed model shown in Fig. 12.7. The shape, which was studied originally by Ahmed *et al.*,[1] is a very simplified vehicle-like form with a hatchback rear, and a backlight angle that can be varied in steps of 5°. It was chosen partly because wind-tunnel data were available, and partly because it provides a difficult test case; the flow field is very sensitive to the separation position on the backlight surface. As was described in Chapter 4, the drag of a hatchback falls initially with increasing backlight angle, but then, as strong vortices start to form over the backlight, the drag rises, reaching a high peak at around 30°–35° backlight angle. Thereafter there is a reduction in the drag once the flow has separated. Test models of broadly similar shape have been used by General Motors, MIRA, Rover, and others.

When using a standard test case for CFD validation purposes, it is of course important to ensure that the experimental data are reliable. In view of the discrepancies reported in the previous chapter for models tested in different wind-tunnels, and the differences between the results obtained with a moving or fixed floor, a large-scale experimental test corroboration programme is necessary before sensible comparison with CFD results can be undertaken. The Ahmed model

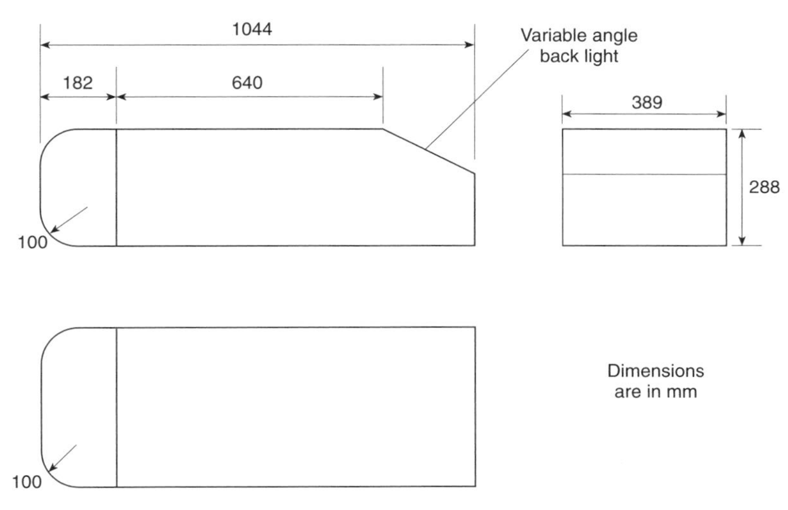

Fig. 12.7 The Ahmed simplified vehicle shape, which is used as a benchmark test for CFD modelling.
(*After Ahmed et al.*[1])

appears to be quite sensitive to flow turbulence, method of mounting, floor clearance and road simulation. As a consequence, different researchers have at times used alternative sets of experimental data.

Several groups have taken up the challenge of applying CFD to the Ahmed shape, including Simcox *et al.*,[6] Graysmith *et al.*[2] and Pearson *et al.*[4] Graysmith *et al.*[2] employed a moving-belt floor for their experimental data, and had the advantage of using a wind-tunnel (the MIRA model tunnel) that has been used in a number of corroborative experimental test programmes. They found that in comparison with experimental data, C_L values were predicted quite well by the computer below a 30° backlight angle, where the flow remains attached, but that above 30° where the flow separates, even the finest grid calculation failed to match the experimental data well. The results for C_D showed a poorer agreement even for the attached flow case, with a 20 per cent discrepancy at 10° backlight angle, rising to 30 per cent at the critical 30°. In general, most researchers have found fairly poor agreement between computed and experimental values for drag coefficient using this model. Centre-line pressure distributions are usually computed well in the front part of the vehicle, where the flow is attached, but tend to be less accurate in the critical area of the backlight, which is unfortunately the source of a large proportion of the drag. Figure 12.8 shows the centre-line velocity vectors and pressure distribution for the 40° backlight case. In this example a coarse grid was used.

Not all shapes are as sensitive to computational details as the Ahmed model, and in Ramnefors *et al.*,[5] for example, a discrepancy of less than 2 per cent in drag

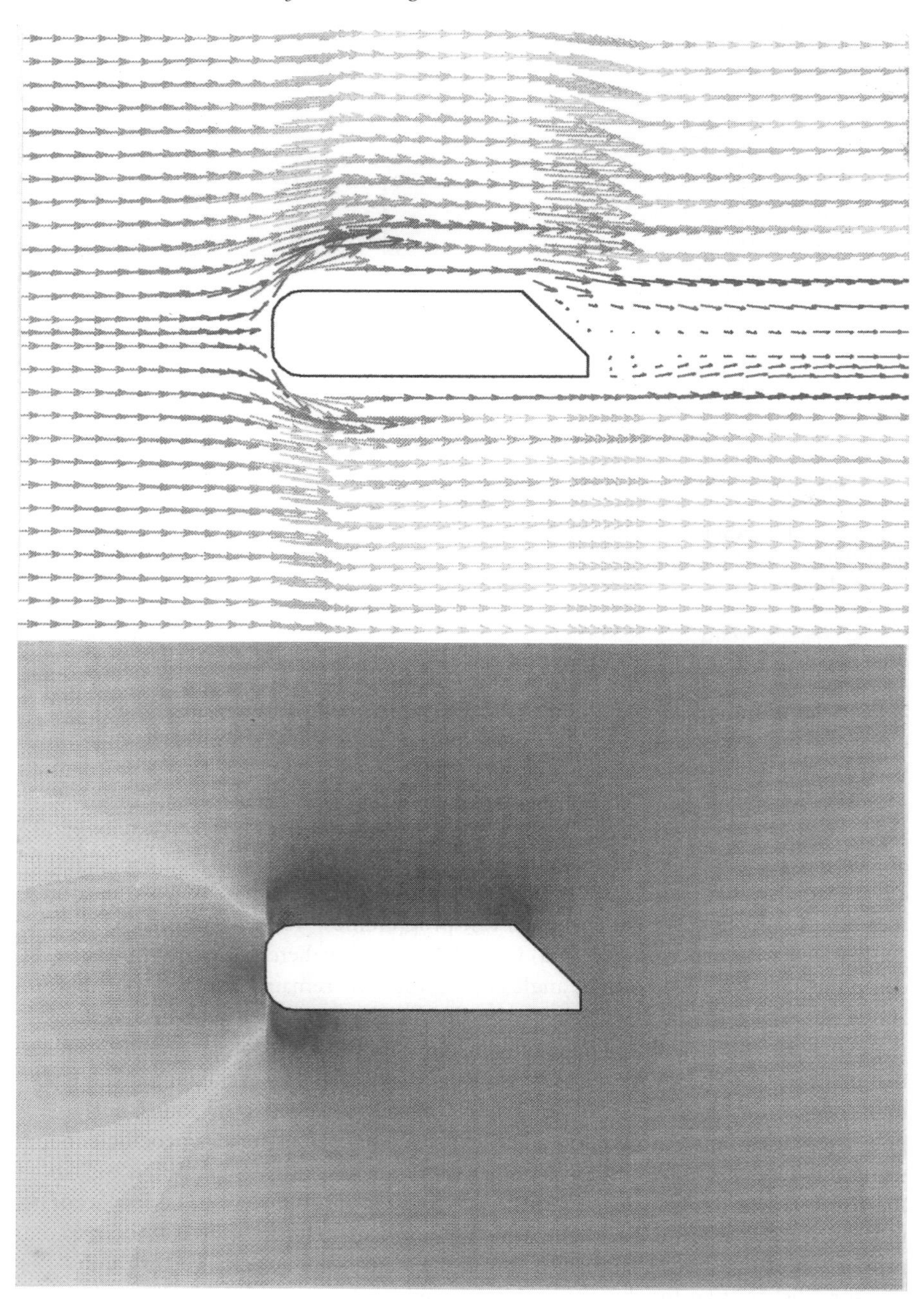

Fig. 12.8 Velocity vectors and pressure contours on the centre-line plane for the Ahmed model with a 40° backlight angle. A coarse grid was used for this example. The contours were colour graded in the original, and much of the definition is therefore lost in this monochrome reproduction. (*Photograph courtesy of Dr S. Wakes, University of Hertfordshire*)

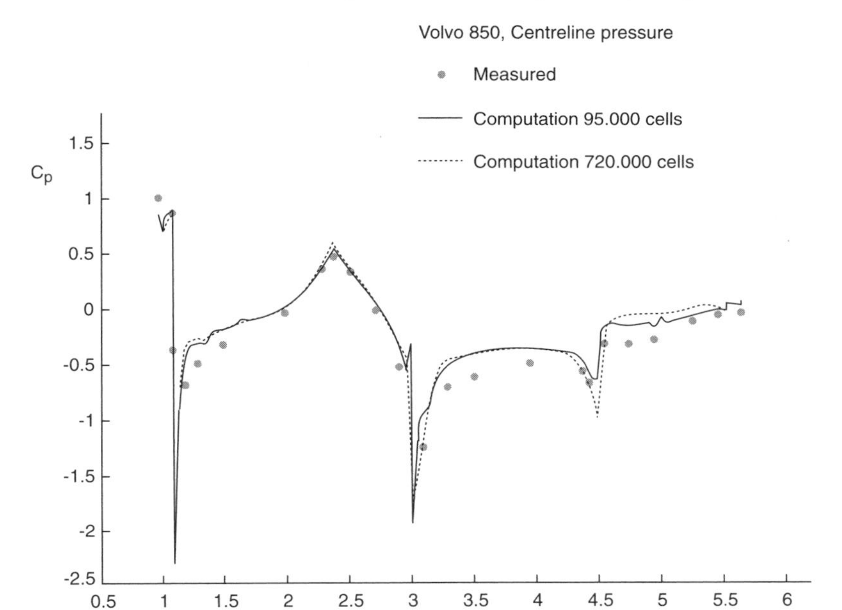

Fig. 12.9 Comparison between computed and experimental data for the centre-line pressure distribution of the Volvo 850.
(*After Ramnefors et al.,*[5] *courtesy of M. Ramnefors and Volvo Data AB*)

calculations was found between CFD and wind-tunnel results for the Volvo 850 when using 164 000 cells. The centre-line pressure distributions are shown in Fig. 12.9. As a result of this study, the authors concluded that with around one million grid cells, the error due purely to grid density becomes small enough for practical engineering purposes.

FACTORS AFFECTING THE ACCURACY OF CFD PREDICTIONS

Various systematic computational tests have been conducted to discover how the different numerical schemes, turbulence models, grid geometry and grid density affect the solutions. In general if all other factors are favourable, increasing the grid density is usually found to improve the match between computed and experimental data, but this is not always the case (see Ramnefors *et al.*[5]), and it is difficult to tell whether a better match is simply due to one set of errors partially cancelling another. Looking at computed overall drag values can be misleading; in Pearson *et al.*,[4] for example, there are cases where an overestimate of the drag on the front of the Ahmed shape is fortuitously partially compensating for an underestimate at the backlight.

It is not only the density of the mesh that matters; the grid shape is also important. Pearson *et al.*[4] found a C_D value of 0.619 for one grid shape and 0.578 for another for the case of the Ahmed model with a 30° backlight angle.

The choice of numerical scheme itself can affect the results. Simcox *et al.*[6] found that for the same test case as above, computed results for drag coefficient varied from 0.455 to 0.354 according to the numerical differencing scheme used.

A final important factor is the turbulence model; computed drag for the the Ahmed shape has been found to be very sensitive to the turbulence model and the choice of parameters used in it (see Pearson *et al.*[4]), particularly at the critical backlight angle.

THE CURRENT RELIABILITY OF CFD CALCULATIONS

It has been found that by juggling with variables such as the grid density, the grid geometry and the parameters in the turbulence model, it is possible to produce computed solutions that do fit the experimental data well. The objection is that as with the earlier inviscid flow calculations, you only get the right answer if you know what it is, so as a predictive tool CFD is still not entirely trustworthy. A further criticism is that different researchers achieve the same answers by different combinations of the various parameters, and sometimes by the fortuitous cancelling of errors. The counter argument is that with the benefit of experience, CFD workers will become increasingly adept at selecting correct values of the controlling parameters.

It has been recognized that the general-purpose CFD package cannot be used blindly as a sledgehammer to crack any kind of nut. Turbulence models that work well on the front part of the vehicle fail in the region of flow separation at the rear, and vice versa. Different parts of the flow field require different treatments, and appropriate boundary and turbulence models have to be applied; also, the grid geometry cannot be selected arbitrarily.

CFD IN THE SHORT TERM

The list of problems outlined earlier might seem daunting, but this does not mean that CFD has no part to play in road vehicle aerodynamics even in the short term. CFD solutions will, for example, identify areas of high and low pressure suitable for siting inlet and outlet ducts. They will also identify areas where separation is likely to take place, or where strong vorticity is being generated. Currently, however, the most useful applications of CFD lie in studying internal flows: cooling and ventilation systems. These can be difficult to model experimentally, and the absolute accuracy required is not particularly high. Computational solutions give a good idea of where the air is going in an internal space, and how fast. Figure 12.10 shows computed values of the velocity vectors in a plane inside the complex geometry of an engine inlet manifold.

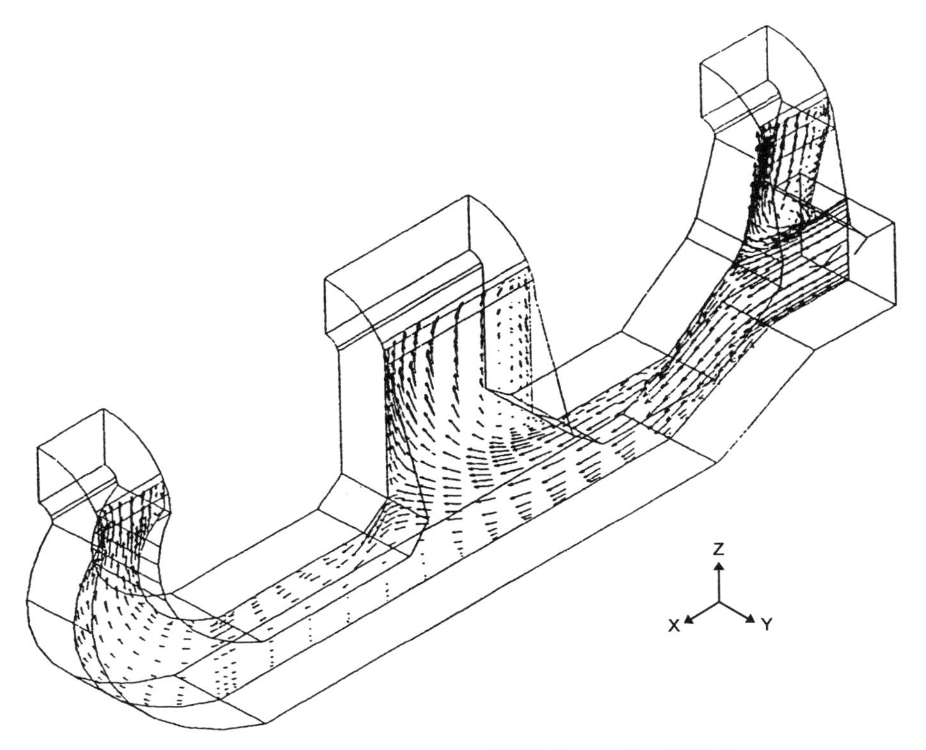

Fig. 12.10 Computed flow field in the inlet manifold of an engine. Velocity vectors on a centre-line plane are plotted.
(Courtesy of Dr K. Mallone, University of Hertfordshire, and Fluid Dynamics International Inc.)

COMPARISON BETWEEN CFD AND WIND-TUNNEL COSTS AND ACCURACY

It should not be thought that computed solutions are cheap; the work station may have a terminal that looks like a personal computer, but the cost will be several orders of magnitude greater. The complex grid shapes take a long time to prepare by skilled operators: typically up to three weeks for a carefully prepared grid of up to 1 million nodes. On very fine grids, one set of results may totally occupy a large computer for a whole week. Each time a small change is made to the geometry, a new run is required, and sometimes even a new grid. On the other hand, if a low order of accuracy is acceptable, then simple coarse grids may be used and the results calculated cheaply on a modest size of work station.

The capital cost of a large well-equipped wind-tunnel is high, but unlike computers, wind-tunnels almost never become obsolete. Wind-tunnel models are expensive, and setting them up may take a few days, but once this has been done, a great deal of high-quality data can be collected in a very short time; one run may take only a matter of seconds. The geometry can be rapidly modified and the model retested immediately.

In summary, for low-accuracy quick results, CFD modelling is relatively cheap, but if large quantities of more reliable data are required, then the wind-tunnel is currently the better source. The discrepancies in wind-tunnel data noted in the previous chapter are generally much smaller than those obtained by different CFD workers. With time, the CFD results will become cheaper and more reliable, but any organization that has a good wind-tunnel is unlikely to swap it for a computer in the near future. Wind-tunnels and computers both have relative advantages, and even in the longer term they are likely to be used as complementary rather than alternative tools.

REFERENCES

1. Ahmed S. R., Ramm G. and Faltin G., Some salient features of the time-averaged ground vehicle wake, SAE 840300, 1984.
2. Graysmith J. J., Baxendale A. J., Howell J. P. and Haines T., Comparison between CFD and experimental results for the Ahmed reference model, *Proc. Vehicle Aerodynamics Conference*, Loughborough, RAeS, 18–19 July 1994, pp. 30.1–11.
3. Katz J., Considerations pertinent to race-car wing design, *Proc. Vehicle Aerodynamics Conference*, Loughborough, RAeS, 18–19 July 1994, pp. 23.1–7.
4. Pearson W. E., Manners A. P. and Passmore M. A., Prediction of the flow around a bluff body in close proximity to the ground, *Proc. Vehicle Aerodynamics Conference*, Loughborough, RAeS, 18–19 July 1994, pp. 29.1–12.
5. Ramnefors M., Perzon S. and Davidsson L., Accuracy in drag predictions on automobiles, *Proc. Vehicle Aerodynamics Conference*, Loughborough, RAeS, 18–19 July 1994, pp. 28.1–14.
6. Simcox S., Jones I. P., Gu C. Y., Ramnefors M. and Svantesson J., The use of Flow-3D for vehicle aerodynamics, *Proc. Vehicle Aerodynamics Conference*, Loughborough, RAeS, 18–19 July 1994, pp. 27.1–11.
7. Stafford L. G., A numerical method for the calculation of the flow around a motor vehicle, Advances in Road Vehicle Aerodynamics 1973, BHRA Fluid Engineering, 1973, pp. 167–83.

BIBLIOGRAPHY

The following books are particularly recommended. Unfortunately one or two of them may not currently be in print, and may therefore be difficult to obtain.

Abbott I. A. and von Doenhoff A. E., *Theory of Wing Sections*, Dover Publications, New York, 1949.

Barnard R. H. and Philpott D. R., *Aircraft Flight*, 2nd edn, Longman, Harlow, 1994.

Benzing E., *Ali/Wings* (in Italian with accompanying English translation) Automobilia, Milan, 1991.

Carroll Smith, *Tune to Win*, Aero Publishers, Fallbrook Calif., USA, 1978.

Hoerner S. F., *Aerodynamic Drag*, Hoerner, PO Box 342, Brick Town N.J. 0873 USA, 1965.

Howard G., *Automobile Aerodynamics*, Osprey, London, 1986.

Hucho W. H. (ed.), *Aerodynamics of Road Vehicles*, Butterworth, London, 1987.

Katz J., *Race Car Aerodynamics*, Robert Bentley, Cambridge Mass., USA, 1995.

Kieselbach R. J. F., Streamline Cars in Europe/USA: Aerodynamics in the Construction of Passenger Vehicles 1900–1945, *Kohlhammer Edition Auto und Verkehr*, Stuttgart, 1982.

Kieselbach R. J. F., *Streamlined Cars in Germany: Aerodynamics in the Construction of Passenger Vehicles 1900–1945*, Kohlhammer Edition Auto und Verkehr, Stuttgart, 1982.

Pankhurst R. C. and Holder D. W., *Wind Tunnel Technique*, Pitman, London, 1965.

Pope A., *Wind Tunnel Testing*, 2nd edn, John Wiley, New York, 1964.

Scibor-Ryski A. J., *Road Vehicle Aerodynamics*, 2nd edn, Pentech Press, London, 1984.

Dorgham M. A. (ed.), *Impact of aerodynamics on vehicle design*, Proc. International Association for Vehicle Design: Technological Advances in Vehicle Design, SP3, 1983.

INDEX